the origins of everything

the ORIGINS of EVERYTHING

EVERYTHING

in 100 Pages (more or less)

David Bercovici

Yale UNIVERSITY PRESS   NEW HAVEN AND LONDON

Yale University Press books may be purchased in quantity for educa-
tional, business, or promotional use. For information, please e-mail
sales.press@yale.edu (U.S. office) or sales@yaleup.co.uk (U.K. office).

Selected images by Casey Reed (frontispiece and illustrations by the
openings of chapters 1–6). The image by the opening of chapter 7 is
courtesy of Frank Fox, http://www.mikro-foto.de; that by the opening
of chapter 8 is courtesy of Denis Finnin, American Museum of Natural
History.

Designed by Nancy Ovedovitz. Set in Quadraat type by Integrated
Publishing Solutions. Printed in the United States of America.

Library of Congress Control Number: 2016938149
ISBN 978-0-300-21513-7 (hardcover : alk. paper)

A catalogue record for this book is available from the British Library.

This paper meets the requirements of ANSI/NISO Z39.48–1992
(Permanence of Paper).

10 9 8 7 6 5 4 3 2 1

CONTENTS

PREFACE

The history of the Universe is perhaps best written backward. Not actually typed that way, but told in reverse chronological order. Our fascination with the moment of creation, whether religious or scientific, comes from our curiosity about how we arrived at where we are now. If we start from the present moment and rewind through 7,000 years of recorded human history, we then find ourselves staring across 7 million more years to the dawn of humanity. As daunting as that seems, it's another 600 million years to the rise of animals, another 3 billion years to the origin of life, and a scant billion more to the birth of our planet and solar system. But from there it's 9 billion more to the dawn of time. If we could play the history of the Universe backward over a 24-hour day, like a brutally long avant-garde movie, then human history would whip by in 4/100 of a second, before the credits had even rolled off the screen; the first animals would appear just 1 bearable hour later; we'd then wait 7 more hours to see the origin of Earth and the solar system, and an agonizing 16 more to reach the beginning of the Universe.

However, as tempting as it is to tell the history of the Universe backward, chronology helps, especially since we're used to thinking and living forward in time. In this brief book, I will, in effect, tell that history in accelerated form, not in 24 hours (though that's up to you, Reader), but in a "quick and dirty" form, with vignettes of major events. This book covers the Universe's greatest hits, recounting when and most importantly how its various pieces emerged. The concept of "origins" is deeply ingrained in science itself: these are not myths or just-so stories but the major scientific origin hypotheses of how things came into existence. And the difference between the just-so story and a hypothesis is critical; researchers can disprove or falsify a hypothesis with experiments or observations because a hypothesis makes a measurable prediction. The testable hypothesis is perhaps the most fundamental precept of science, and although it might sound a little dry, I hope to relay a flavor of this through these origin stories. Don't worry; I won't flavor it too much.

I'll note that this book arose from a class, an undergraduate seminar at Yale with the modest title "Origins of Everything," whose goal was to teach science through these "big" testable hypotheses. While the book's material is geared toward a general audience, I don't believe in going too fluffy on the science. However, I will also do my best to avoid bamboozling the reader with jargon, and I'll try to explain it if and when a little jargon is necessary.

Although I'm offering vignettes of origin stories, they're not random or disconnected; each one depends on the last and flows into the next. The building blocks of life are drawn from the air, sea, and rocks of our planet, which itself grew from interstellar dust. The elements in that dust were forged in giant stars, which were themselves born from gas created in the Big Bang. Where the planet is, how its oceans, atmosphere, and deep interior formed and change are all responsible for sustaining complex life for hundreds of millions of years.

Of course, as a scientist who has researched a number (though certainly not all) of the topics covered here, I'll invariably emphasize themes and connections between the origin stories from a uniquely geophysicist's perspective—or, more honestly, bias. As my students eventually figure out, plate tectonics plays a formidable role in this narrative, and if I could figure out how to make it responsible for the Big Bang, I would (but there's an annoying problem with timing). There are other excellent books, far more comprehensive than this one, on the history of the Universe and life, and I list and recommend them all at the end. The goal of this book is not to be deep and comprehensive but instead to be boldly (or baldly) shallow and superficial, in the best sense of these words . . . if there is one. My aim is to give a quick and, I hope, readable overview that provides a taste of our Universe's story (and to some extent humanity's place in this story) and more importantly to give you, the reader, an appetite to learn more.

Disclaimer: Given the scope of this short book, one might be misled into thinking that I'm an expert on all of its subjects. While that would be wonderful if true, frankly, it's not. My own knowledge comes from covering all these subjects at some level over nearly 30 years of university teaching, but I am certainly not an astronomer or a biologist or an anthropologist. In the same regard, subjects that come closer to my own field of geophysics and planetary science surely receive more detail. Therefore, the reader should not rely on this book as the definitive word on the wide number of subjects that we'll skim through. This book is more like a sampler plate at a fusion restaurant whose chef is best known for linguini.

the origins of everything

1 UNIVERSE AND GALAXIES

Time begins with an unfathomably colossal explosion, which is always a good way to start a story. However, whether that first moment marked the creation of the Universe or the creation of Earth was unknown until fairly recently, within the last century. Indeed, the first words in the Judeo-Christian Bible are "In the beginning God created the heavens and the earth," a single event that was dated by the seventeenth-century Irish bishop James Ussher to have occurred exactly on October 23, 4004, BCE.

But not so long before Ussher's work, various prominent Renaissance philosophers held the radical view that time had no beginning. One of the most famous of them, largely because of his martyrdom, was Giordano Bruno, a sixteenth-century Italian Dominican monk and scholar. Bruno supported Copernicus's then controversial theory that the Earth was not the center of the Universe but instead orbited the Sun. He went further to propose that the Sun itself was just a star like many others in the night sky, all with planets of their own. But most significantly (for our story, anyway), Bruno believed that the Universe was unchanging and of infinite age and extent. Although Bruno was not the first scholar in Europe to hold such views, the Catholic Church declared them heresy (along with his more religiously offensive notions, for example, about the divinity of Jesus Christ and the validity of transubstantiation). He was eventually captured in Venice, tried, extradited to Rome, and tried again. Bruno was hotheaded and sarcastic, and in true form, he refused to recant his writings unless the pope or God himself told him he was wrong. Neither consented, and Bruno was burned at the stake on Ash Wednesday 1600 in Rome's Campo de' Fiori, where today a statue of him glowers at the cheerful tourists in the plaza's many cafés.

Thankfully, scientists are rarely burned at the stake anymore (at least not literally) for their ideas. When I was visiting Rome once, a colleague and I stood below Bruno's imposing statue and pondered whether we would recant our scientific writings under pain of death, as Galileo Galilei did 33 years after Bruno's trial. After a brief pause, and admittedly a burst of laughter, we agreed that of course we would, in a heartbeat. Our mutual cowardice aside—and the thought of dying for all our papers that no one has read—we have the vantage of hindsight and know that bad science dies with its progenitors, but good science does not. If our scientific views died with us, then that is probably what they deserved. But Bruno did sacrifice his life for his beliefs, becoming one of our most renowned martyrs of science. And, in the end, his views were recognized as surprisingly prescient, especially his idea that Earth is just one of many worlds orbiting around one of many stars in a vast and ancient Universe.

Bruno's views that the Universe is infinite in age and extent are, however, wrong, and time does indeed have a beginning. The simplest evidence for this fact is the darkness of the night sky. If we lived in an infinitely old and infinitely large Universe, then every patch of the night sky would have a star somewhere in it, and light from all those stars would have had ample time to reach us; thus the whole sky should be illuminated with starlight. Although the German mathematician Johannes Kepler and English scholar Thomas Digges (contemporaries of Bruno) noted the nighttime sky paradox, it is named after the much later eighteenth- and nineteenth-century German astronomer Heinrich Wilhelm Olbers. The solution to Olbers' Paradox was found by the nineteenth- and twentieth-century British physicist William Thomson, aka Lord Kelvin, and even inferred by the American author Edgar Allan Poe: that is, the Universe is either of finite age (thus light from distant stars might not have reached us yet), or of finite extent (not every patch of sky has a star in it), or both. This is one of the first and most important clues that eventually led to the Big Bang hypothesis, since it implies the Universe started at some moment in the past and/or not everywhere all at once.

In the 1920s, the American astronomer Edwin Hubble used telescopic observations to show that there are galaxies beyond our own Milky Way, which was originally thought to be the extent of our finite and static Universe. Hubble inferred the distance to galaxies using pulsating stars, called Cepheid variables, whose pulsation period (time between pulses) and average luminosity (total power released as light) are related in a simple way. This feature makes them good landmarks for gauging distance: two Cepheids with the same pulsation have the same luminosity; thus, if one of them appears dimmer than the other, then it is farther away, and measurably so (since the dimming effect is simply related to the square of its distance). Thus, the Cepheids give the distance to the galaxies in which they are embedded. Hubble also found that, on average, the light from galaxies appears increasingly "red-shifted" the farther away the galaxies are. Red light has the longest period and wavelength of visible light waves. Red shifting of light is similar to how sound waves emitted by an ambulance siren shift to a lower pitch (to lower frequencies, or longer periods and wavelengths) as the siren moves away from you. Redshifted light from galaxies indicates that the farther away they are, the faster they are moving away, meaning galaxies are generally all moving away from each other and spreading outward.

Even before Hubble's observations that the galaxies are, on the whole, spreading away from each other, the Belgian astronomer Georges Lemaître and the Russian physicist and mathematician Alexander Friedmann independently predicted that the Universe is expanding. Both used Einstein's general theory of relativity for their calculations, although Einstein himself originally rejected their conclusions (he accepted them later). But Hubble's observations provided the evidence to support their ideas of an expanding Universe.

If the Universe is of finite age and extent and is expanding, then all its mass and energy were originally—playing the expansion backward—concentrated in a much smaller and hotter volume, what Lemaître called a "cosmic egg." The initial expansion of this mass, at the first moment of the Universe, was disparagingly coined the Big Bang by Cambridge astronomer Fred Hoyle, who actually hated the idea. The name stuck, although the term *bang* suggests an explosion, which is not really correct—despite my opening sentence of this chapter. An explosion involves a shock wave separating a high-pressure gas from a low-pressure one, while the originally compacted Universe's mass and energy and hence all space were inside its tiny volume; thus, there was nothing to expand into. As the Universe expands it carries the edge of space with it; outside this boundary there is no light, matter, energy, or time—a concept that is hardly intuitive.

Finally, in the 1960s, Americans Arno Penzias and Robert Wilson discovered the Cosmic Microwave Background (CMB) radiation—a radio hiss that now pervades the Universe—which showed that the deep vacuum of space is not dead cold, with zero temperature and energy, but is filled with microwaves at a "balmy" 3 degrees Kelvin (−270 degrees Celsius). This lingering warmth is evidence of the hotter state of the Universe following the Big Bang.

The Big Bang theory, or really just using current observation of the Universe's expansion, allows a simple prediction of the age of the Universe. If we calculate the amount of time it would take the Universe to grow at its estimated expansion rate (called the Hubble constant) from a small volume to its current size, and cool so that space is at a temperature of 3 Kelvin, we would predict that the age of the Universe is about 14 billion years (plus or minus a billion years or so). This simple inference is fairly well verified by astronomical observations of the Universe's oldest objects, which are typically small stars that burn very slowly (as we'll discuss in the next chapter), although these would not have been born until a few hundred million years after the Big Bang, so they're an underestimate of the Universe's age. The current best estimate of the age of the Universe is 13.8 billion years.

The Big Bang theory, however, is much more than just a description of the growth of the Universe from a tiny point to its current vast size. The sequence of events after its initial state governs the very nature of matter and the structure of the Universe. In short, a lot happened in just the first few tiniest fractions of a microsecond up to about a minute after the Big Bang itself. In general (before we get into the weeds), we can think of the initial Universe being so condensed and hot that it was a very tiny ball of pure, immense energy, and as it expanded and cooled, the various states of matter, energy, and even forces of nature condensed out, the way we might loosely think of steam cooling first from gas to liquid water and then from water to solid ice. Each step results in a different state of matter (gas, liquid, or solid) and is called a phase transition. But these transitions in the first few moments of creation were far more exotic, and they started from an initial state that we have yet to fully understand.

At the very first moment, just at the start of the Big Bang, it's thought that temperatures and, in effect, pressures were so high that the Universe (such as it was) contained only one form of extreme energy, existing in an unimaginably tiny volume, much, much smaller than an atom or even a subatomic particle. This state existed for the first 10^{-43} seconds. (For future reference, 10^{-2} is the same as 0.01 and thus 10^{-43} is a 1 separated from the decimal point by forty-two zeroes.) This stretch of time is called the Planck epoch, after Max Planck, the twentieth-century German physicist who is widely recognized as the father of quantum mechanics. During this epoch (and I'll note here that cosmologists employ a whimsical use of the terms *epoch* and *era* that would drive most geologists crazy), the basic forces of nature were also just one force. Specifically, forces involve exchanges of particles; for example, kitchen magnets stick to your refrigerator by exchanging particles called photons, which are both electromagnetic "force carriers" and particles of light. Other forces now have different force carriers, but if all force-carrying particles were the same in the Planck epoch, then all forces would have been the same. The concept of an initially single force is the theoretical physicists' long-sought "Unified Field Theory" or "Theory of Everything." Currently, a theory explaining how to unify gravity (which holds us on this planet) with the other three fundamental forces—electromagnetic (which controls the force between electric charges and also includes magnetic forces), strong, and weak forces (which control how subatomic particles are held inside atomic nuclei)—is elusive. Entire fields of physics like string theory or loop quantum gravity are attempts to crack this difficult nut. Unifying the three forces other than gravity has reached more success both theoretically and experimentally; this is called the Grand Unified Theory, which falls within what is called the Standard Model for "almost everything"—that is, everything except gravity. The discovery of the Higgs particle or boson (named for the British physicist Peter Higgs) provided a groundbreaking step and validation for the Standard Model, and in principle explains what gives matter the property of mass (in that "inertial mass"—what makes some objects harder to get moving than others—is really due to the drag on particles by a ubiquitous Higgs field).

But I digress. The real point is that we still have little idea what the state of the Universe was during the Planck epoch itself, nor how it got to that point in the first place, or what came before. Regardless, at the end of the Planck epoch, the tightly bound tiny Universe was unstable and the Big Bang started.

The next 10^{-35} seconds of the Universe could really be called the bang of the Big Bang, involving extremely rapid inflation; this minuscule stretch of time is called the Inflationary epoch. Inflation would have expanded the Universe in volume by many, many factors of 10 (like 10^{70}), and while this would have brought the size of the Universe only to a relatively smallish size (say a few meters), it would have occurred many times faster than the speed of light. The inflation is thought to have been driven by the release of just one form of energy stored in a single force field, and that burst of energy became the source of matter and energy of the known Universe that followed.

The reason rapid inflation has become an integral part of the Big Bang story is that without it, the basic observation of the CMB radiation, that pervasive radio hiss mentioned earlier, would be hard to explain. For example, since all empty space in the Universe seems to be at nearly the same temperature after nearly 14 billion years, distant different ends of the Universe must have been in contact with each other for long enough and up to a large enough size that they carried the same temperature with them afterward. If they were never in contact with each other from time zero, then it's hard to understand why they would all be at the same temperature now. Rapid inflation allows the Universe to have quickly reached a small but finite volume in which all parts of the Universe could attain the same temperature, before those parts blew apart in all directions.

After inflation, the released energy expanded to lower density just enough to condense out matter. Energy can be converted to matter according to one of the few of Einstein's equations that most people remember, $E = mc^2$, where E is energy, m is the converted mass, and c is the speed of light. This first matter appeared mostly in the form of a soup of sub-subatomic particles called quarks, which are the building blocks of protons and neutrons, which are in turn the building blocks of atomic nuclei. There was also still a huge amount of pure energy in the form of photons as well as a group of much less massive particles of matter generically called leptons. Leptons include electrons, the tiny negatively charged particles orbiting atomic nuclei and carrying electrical currents in wires, as well as nearly massless neutral particles called neutrinos, which are zipping unnoticed through your body right now. In general, leptons are defined in part by the fact that they can't combine together to make atomic nuclei.

At this point in the story, temperatures were still initially too high to allow quarks to combine with each other. But in the next 10^{-5} seconds, a lot more happened. Matter and what's called antimatter existed in nearly equal amounts; for example, the opposite or antimatter version of an electron is a positron with the same mass but opposite electrical charge. But matter and antimatter annihilate each other on contact, and so after a brief moment of coexistence, this is what they did. This annihilation released more energy but left behind a "little" regular matter, which was slightly more abundant, and thus exists now. Dark Matter, which is thought to be most of the mass of the Universe (more on this below), was also likely created during this time. The last stages of this stretch of time involved the combination of quarks, now cool enough to link into protons and neutrons. But conditions were still too hot to join neutrons and protons into atomic nuclei, let alone make complete atoms. Nevertheless, the last fraction of this 10^{-5} seconds is called the Hadron epoch, because protons and neutrons are generically called hadrons.

After these first 10^{-5} seconds, temperatures remained high enough and photons were still sufficiently energetic that they could convert their energy into matter and continue to make leptons. But after 1 second, conditions had cooled enough to stop lepton production, thus more or less leaving the amount of leptons we have today (except for those created in nuclear reactions), so that 10^{-5} to 1 second is called the Lepton epoch.

After about 1 second up to about 100 seconds, the Universe was cool enough that neutrons and protons could link up to make the first atomic nuclei. But free neutrons on their own are not stable and tend to decay into an electron and a proton. Thus, by the time 100 seconds were up, not many neutrons were left: out of every sixteen hadrons, only two were neutrons while fourteen were protons. In this batch of sixteen hadrons, two neutrons would combine with two protons to make a helium nucleus, and leave twelve protons, each of which made a hydrogen nucleus. Thus, about a quarter of the normal mass of the Universe was helium (that is, four out of every sixteen hadrons were in helium) and the rest was hydrogen (twelve out of every sixteen hadrons were hydrogen). In fact, a little bit of other matter, like lithium and heavier species (isotopes) of hydrogen (such as deuterium, which has both a neutron and a proton in its nucleus), was also made but in tiny quantities since conditions had cooled too rapidly to form more of them. That the Universe today is still filled with this mass fraction of about 75 percent hydrogen and 25 percent helium, and a smattering of heavier elements (more on this below), is a further testable prediction (and successfully tested at that) of the Big Bang theory.

For the next 100,000 years, the Universe was still too hot to allow atomic nuclei to capture electrons and form whole atoms. The density of both matter and energy from photons was high enough that they were stuck to each other. This means matter was thick enough to be opaque and energy high enough to keep matter from clumping into anything but separate nuclei and electrons; this is often called the

Radiation epoch since the Universe was bathed in photons. At about 100,000 years, both mass and photon density decreased enough that light could escape. By about 380,000 years, the temperature fell enough so that nuclei and electrons could combine to make atoms, thus leading to the Matter epoch we are basically still in today. That last combination released a burst of energy, the remnant of which is the CMB glow. This final nuclear combination and release of energy would carry signs of the slightly lumpy uniformity of the quark soup following rapid inflation; thus, the uniform but slightly speckled pattern of the CMB now is thought to reflect the hugely expanded ghost of this first fingerprint.

After light's escape following the Radiation epoch and the burst of energy from atomic combination, the Universe fell into blackness for the next 300 million years, in what is called the Dark Ages. In short, the Universe's temperature had cooled enough and matter was dispersed enough that there were no sources of light.

At the end of the Dark Ages, the weak fluctuations in density in the hydrogen-helium gas caused gravitational attraction toward the denser regions, which thus drew in more matter. The additional matter caused these fluctuations to get denser, draw in more mass, and so on, to create the first gravitationally bound structures in the form of vast clouds. Within these clouds of gas the first stars were born.

The first stars to form would be made only of hydrogen and helium, and their formation marked the end of the Dark Ages, 300 million years after the Big Bang. After the biggest of these first stars came and went and created heavier matter (more on this in the next chapter), other, smaller stars would form in droves within giant clouds and then be gravitationally bound together to make the first real galaxies, whose production peaked by about 1 to 3 billion years after the Big Bang. Though on average the galaxies in the Universe are expanding away from each other, they are not entirely free floating, and groups of them are gravitationally bound to each other to form clusters of galaxies. These clusters are themselves bound to each other along filaments that permeate the Universe. Webs of these filaments are the largest structures in the Universe, and in between the filaments are voids of space.

Our own galaxy, the Milky Way, is bound to the Andromeda Galaxy (they may even collide in the distant future), both of which are large galaxies in the Virgo Cluster (which may itself be part of an even larger cluster named Laniakea). Nevertheless, after the formation of the first galaxies a billion years after the Big Bang, it likely took another 1 or 2 billion years for these galactic clusters and filaments to organize.

Galaxies today are not uniform in size or shape, but neither are they entirely random in contour. The largest ones are elliptic galaxies that are more or less spherical balls of stars, all orbiting the center with randomly oriented orbits. Fairly common ones are disk, spiral, and bar-armed galaxies that are flat and appear to be spinning about a central mass—for example, the Milky Way and Andromeda galaxies. Indeed, as a giant cloud of gas and stars that formed a spinning galaxy would have collapsed, its spin would have inhibited collapse perpendicular to the spin axis but freely allowed collapse parallel to this axis, thus creating a flat disk shape (similar to solar system formation, which we'll discuss later). As with solar systems, the center of the collapsing cloud usually gets most of the mass; for a solar system this would be the star. For the galaxy, the central mass is so huge it forms a supermassive black hole, a mass so large and dense that light cannot escape its gravitational attraction if it gets too close.

Galaxies are typically some 100,000 light years across. (1 light year is the distance traveled by light in 1 year, which is about 10^{13} kilometers, or 10 trillion kilometers; by comparison, Neptune, our farthest planet, is about 4.5 billion kilometers away from the Sun, a distance about two thousand times smaller than 1 light year.) Our galaxy contains hundreds of billions of stars. However, several lines of evidence suggest that the observed stellar mass of galaxies is only a tiny fraction of the total galactic mass; there is an immense amount of invisible mass within galaxies that is aptly called Dark Matter.

In the 1960s, the American astronomer Vera Rubin and colleagues discovered that most stars in the spirals and bar-arms of disk-shaped galaxies orbit their galactic center with nearly uniform speed regardless of distance from the galactic center, which is very different from how our planets orbit the Sun. The planets' orbital speed decreases with the distance from the Sun because the only force holding them in orbit is the gravitational attraction of the Sun, which gets weaker with increasing distance (in what are called Keplerian orbits, named for Johannes Kepler and his laws of planetary motion). The uniform orbital velocity of stars means that the further they are from the galactic center, the more mass there is inside their orbit that is gravitationally binding them to the galaxy. However, the total mass required to keep stars orbiting this way is much more than what can be seen of normal galactic mass, suggesting the presence of Dark Matter to make up the rest of the necessary mass.

Astronomers also observed that the relative velocities of galaxies within galactic clusters are too fast to stay gravitationally bound to each other if their mass is comprised only of the observed stellar mass; that is, the clusters are stable and don't fly apart only if there is much more mass holding them together than can be seen. There have been various other lines of evidence for Dark Matter as well, for example, gravitational lensing, in which light is bent as it passes by massive objects like galactic clusters.

This invisible Dark Matter holding galaxies and clusters together is undetectable in all wavelengths of light (from microwaves to infrared to ultraviolet). However, in recent years, astronomers have had to conclude that a huge fraction of matter in the Universe is dark, and that the first galaxies were made more of Dark Matter than of hydrogen and helium. The fundamental composition of Dark Matter remains a mystery since its presence is detected only indirectly.

Since the Universe is still expanding outward from the Big Bang, one naturally wonders about the future—whether its expansion is slowing under the pull of gravity but has enough of the initial explosive energy to keep going, or whether it will "run out of steam" and gravity will cause the Universe to collapse back toward its center. Recent discoveries have implied that neither scenario is correct, but that the expansion is accelerating. Gravity was, until these discoveries, the lone long-range force, and it can only cause deceleration (and possibly collapse) of the Universe under the attraction of mass. This accelerated expansion was thus a big surprise and gave evidence of another heretofore undetected force, fueled by an energy field called Dark Energy, which essentially provides the pressure to inflate the Universe faster. (That both Dark Matter and Dark Energy are termed "dark" doesn't mean they're related; they're just both impossible to see using light.) Dark Energy is an ultra-long-range force that acts only across galactic-super-cluster scales, and may not have become important until the Universe expanded to a large enough size. The dominance of Dark Energy over gravity, and the resulting accelerated expansion, is inferred to have happened about 4 billion years ago, even after our solar system formed. In a way, this expansion is analogous to the Universe filling up a gradually sloping basin but then cresting a rim and spilling down the other side.

Given the volume of the Universe over which Dark Energy extends, it is inferred to be most of the Universe's bulk (mass and energy combined) at about 70 percent, while Dark Matter makes up about 25 percent; the remaining 5 percent is normal atomic matter, from which stars, planets, and you and I arise, although most of it is still hydrogen and some helium. But Dark Matter and Dark Energy make themselves known only over the scales of galaxies and galactic clusters, which are not distances for which we have any feeling, concrete experience, or intuition. Gravity is pretty much the only force we physically sense and navigate on a regular basis— getting out of bed, climbing stairs, pouring coffee. However, if we were the size of a small bug or a microbe, our life would be dominated as much or more by electromagnetic forces that cause static electricity and surface tension on water, and we would find gravity less consequential or barely noticeable (for example, an ant is hardly impeded climbing walls and would be unaffected by falling off a tall building); thus, as far as Dark Matter and Energy are concerned, we're down at that microscopic bug scale.

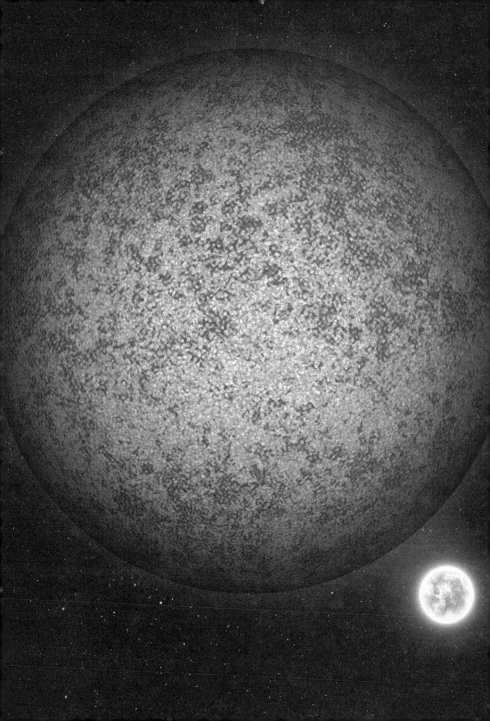

2 STARS AND ELEMENTS

The Dark Ages of the early Universe ended when vast clouds of hydrogen and helium (and Dark Matter) collapsed under their own gravitational self-attraction and began to form the first stars and then galaxies. Similar star formation occurs today, and one example is the Eagle Nebula in our own galaxy, which is still churning out new stars and solar systems. But as we noted before, the first such gas clouds contained only atomic matter (that is, aside from their Dark Matter) in the form of mostly hydrogen and helium, and so there was nothing yet with which to make planets. The formation of these first and subsequent stars built heavier elements from which the planets and things living on them were created.

Once a pre-solar cloud starts to collapse under its own gravitational self-attraction, its molecules fall toward its center and pick up speed (much like a ball rolling down a hill). The accelerating molecules collide and bounce off each other, and the energy of their motion is converted into heat; this raises the cloud's temperature and pressure, which then halts its collapse. (We'll discuss more about the size and shape and evolution of these clouds in the next chapter.)

In fact, the cloud may not collapse very far on its own before it stalls, depending on its size. If it's not too massive, then it won't collapse much at all; but with larger mass, gravity is stronger and thus the cloud collapses even further before it becomes too hot.

Some processes help sustain the cloud's collapse. For example, most of the cloud is made of hydrogen gas whose molecules are composed of two hydrogen atoms bound together. If the center of the collapsing cloud gets hot enough for the hydrogen molecules to break apart into atoms, then that separation soaks up energy and keeps the temperature from increasing, allowing the cloud to keep going. This process is similar to the phase change that occurs in boiling water (we used a comparable analogy in describing the Big Bang). Pumping energy from your stove into water heats it up until it starts boiling. But the conversion from liquid to gas during boiling soaks up energy, and the temperature stays the same until all the water is boiled away. Thus, the conversion from molecular to atomic hydrogen soaks up heat from the collapsing cloud and keeps its temperature stable until the conversion (in that part of the cloud that got hot enough to drive this conversion) is finished. A similar step occurs later on or deeper inside the cloud, where it's even hotter yet, when the temperatures are high enough to strip electrons from hydrogen atoms and ionize them. That process acts like another "phase change" that levels off temperature.

Even so, only extremely large clouds are able to collapse enough entirely on their own without any help. Thus, the very first stars that were made up entirely of hydrogen and helium were massive (these are traditionally called Population III stars, which today are elusive beasts), starting with clouds that were thousands or even millions of times more massive than our Sun, and eventually forming stars hundreds of times more massive than our own. For small clouds leading to smaller stars, the collapse needs a trigger and a push to get it over the threshold beyond which it is dense enough to keep contracting. Giant stars, for example, often die in explosive supernovae (more on this below) and produce shock waves that can whack a neighboring cloud and get it to start collapsing. Help like this was likely needed to start the formation of the first small stars, which live a very long time and yield some of the main evidence for the age of the Universe. There is evidence in meteorite dust grains that our own solar system was started this way, but we'll loop back to that a little further on.

If all the conditions work out to allow a collapsing cloud's peak temperature to reach about 10 million degrees Celsius, then a star will be born. At this temperature, the nuclei of ionized hydrogen move fast enough to overcome their electrical repulsion (they're just protons at this point, each with positive charge, and thus repel each other) and fuse into helium, whose nuclei are typically made of two protons and two neutrons. This nuclear fusion process gives off an enormous amount of energy because of the conversion of mass into energy. As discussed in the previous chapter, Einstein's most famous equation for conversion of matter of mass m into energy E states that $E = mc^2$, where c is the speed of light, equal to about 300,000 kilometers per second (a speed that would allow you to go around the circumference of Earth about eight times in 1 second). Since c^2 is such a huge number, it means that converting a mass of just 1 milligram (about the mass of a very small pill) into energy is enough to boil away about 40,000 liters of water, or alternatively, converting 60 milligrams (a very small bottle of tiny pills) would vaporize an Olympic-sized swimming pool. The process of nuclear fusion was discovered during the 1920s and 1930s, and subsequently used to develop the theory of stellar nucleosynthesis (most notably by physicist Hans Bethe and astrophysicist Fred Hoyle, although predicted much earlier by astronomer Arthur Eddington), which is described here.

In the collapsing pre-solar cloud, this first conversion of matter to energy occurs because the mass of four hydrogen atoms is very slightly greater than the mass of one helium atom, and much of the leftover unnecessary mass is turned into energy. This massive release of heat stops further collapse of the cloud, and holds the peak temperature at just over 10 million degrees Celsius (in the Sun's core, it's about 15 million degrees Celsius). This stalled cloud is in fact a star, just as the Sun is now: a condensed cloud of gas that has stopped collapsing because of the heat released from nuclear fusion.

This fusion reaction does not occur everywhere inside the star, but only in its deepest, hottest layer, called the core. Outside the core, temperatures are too cool to drive fusion, and heat from the core is carried toward the surface by convection—wherein hot buoyant stuff rises, leading to the Sun's grainy appearance, called solar granulation—and away from the Sun by radiation or photons, which eventually reach Earth as solar energy in the form of visible light. The Sun also blows off heavier particles like loose electrons and protons that travel out in the solar wind, which eventually hits the Earth and other planets.

Stars as small as Sol, our Sun, or smaller ones like red dwarfs, have their collapse halted just with the "moderate" temperature maintained by hydrogen fusion. Although they are small, they will maintain this state of "burning" hydrogen for a very long time. The reason for this slow burn is that the process of fusing hydrogen atoms is itself infrequent, since it is essentially impossible to get four hydrogen nuclei, or protons, to stick together at once to make one helium nucleus. Thus, the process involves several steps in what is called the proton-proton reaction sequence. In this case, the first two protons slam together, overcoming their electrical repulsion, and join briefly to make a two-proton nucleus that is a light isotope of helium. (Different isotopes of a given element have atomic nuclei with the same number of protons but different numbers of neutrons, which, being neutral, don't affect the element's chemical properties; all helium isotopes have two protons but can have anywhere from zero to eight extra neutrons, although only the ones with one and two neutrons are stable—that is, don't decay to something else.) This light helium nucleus is unstable and doesn't last long. It kicks out some antimatter, the positron or an anti-electron, as well as a tiny neutrino (which then accounts for the solar neutrino flux) in order to turn one of its protons into a neutron, leaving an isotope of hydrogen called deuterium, with one proton and one neutron. This deuterium is then hit by and fuses with a third proton (again, a hydrogen nucleus) to make another isotope of helium, with two protons and a neutron, which is stable. The final step is that two of these types of helium nuclei slam into each other and make yet another form of stable helium, with two neutrons and two protons, and kick out two protons. This last step releases much of the fusion energy, and the two rapidly ejected protons continue on to smash into other protons and perpetuate the slow chain reaction. (This final helium nucleus, with two protons and two neutrons, is also called an alpha particle, and it is often the product of nuclear decay of much heavier atoms, like uranium.)

I admit, that's a fair bit of detail about the hydrogen fusion reaction, but it deserves our attention for two important reasons. First, the fusion reaction is the basic energy source that powers the Sun and thus life on Earth. It is also the power source for the activity of the oceans and atmosphere, including sea currents, weather patterns, and even climate variations. Second, the proton-proton reaction is so slow that in the Sun's case, the hydrogen burn lasts about 10 billion years, a period we are about halfway through now; this is a good thing, really, since it's taken a big chunk of that time for life on Earth to evolve to complex forms like us. But apart from that, a small star like ours is pretty useless for making the building blocks of planets, since it is only making new helium, which already existed anyway from the Big Bang. So stars like the Sun are nothing special, at least as far as creating new matter is concerned.

Much larger stars, fifteen or more times the mass of our star, won't stop their initial collapse at the paltry temperatures of 10 to 15 million degrees Celsius. Their contraction will stall only at much higher temperatures, which, if high enough, will form even heavier elements by fusion. For example, at about 100 million degrees Celsius, stars can fuse helium to make carbon and then oxygen. With very large stars, called red supergiants, temperatures are so hot that they can form elements all the way up to iron.

Some of the most important fusion processes for making heavier elements involve combining the helium nuclei called alpha particles (which, as noted above, each have two neutrons and two protons). One example is called the triple-alpha process, which entails two reactions that bring together three alpha particles to make carbon; this is a difficult and infrequent reaction and thus is a bottleneck for forming heavier elements. But once carbon is present, the "alpha-chain" process takes over and sequentially adds one alpha particle at a time, starting with carbon to make oxygen, then neon, then magnesium, then silicon, and so on up to iron (although it actually first makes unstable nickel, which then decays radioactively to stable iron). Each such step happens only at a much higher temperature and pressure than the previous step, so they each tend to occur at deeper and hotter layers inside a giant star, which is thus layered like a big onion, each deeper layer a factory for heavier elements.

The uppermost layer of this stellar onion that is still hot enough to sustain fusion turns hydrogen to helium, and thus it supplies most of the feeder material for all the layers below it. If this layer created helium as slowly as the Sun did, then it would choke off the factories below it or at least be a major bottleneck, since the reactions at greater temperature and pressure happen much faster and would thus use up the helium trickle quickly. But the fusion of hydrogen into helium is accelerated in these stars by the presence of carbon, nitrogen, and oxygen in what is aptly called the CNO reaction, thus rapidly producing the helium or alpha particles that are used at greater depth.

The deepest, hottest layer, near the center of the star, can, if hot enough, make stable iron (by way of unstable nickel), but that's where the fusion stops. To make any element heavier than iron involves creating mass, since the element being created has more mass than its components; thus more energy would be soaked up to create the new element rather than be released. Instead of powering the reaction further, such fusion would cool it off and stop it.

Some of the most abundant elements in our solar system (after hydrogen and

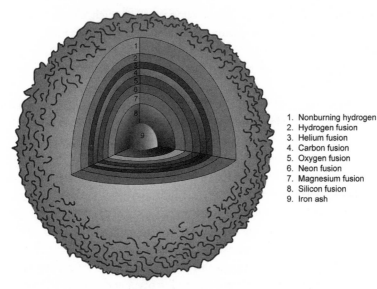

1. Nonburning hydrogen
2. Hydrogen fusion
3. Helium fusion
4. Carbon fusion
5. Oxygen fusion
6. Neon fusion
7. Magnesium fusion
8. Silicon fusion
9. Iron ash

Giant stars are structured more or less like onions. Each deeper, hotter layer is a factory for fusing lighter elements from the overlying layers into progressively heavier elements, such as hydrogen into helium and helium into carbon, all the way up to iron. Many of the reactions involve fusion with helium nuclei in the "alpha" processes, thus creating some of the elements necessary to make planets and life, such as carbon, oxygen, magnesium, silicon, and so on. (Courtesy Barbara Schoeberl, Animated Earth LLC.)

helium) are made up entirely of these alpha particles, and these are most of the building blocks of life and planets, including carbon, oxygen, silicon, magnesium, calcium, and iron. This gives us one possible reason why life is carbon based. As one of the first stable elements in the alpha production chain, there's a lot of it. Carbon is also an enormously versatile atom; it makes many kinds of compounds or chemicals, especially with the ubiquitous hydrogen, leading to organic molecules that are the building blocks for life. Other important elements, especially for life, like nitrogen and phosphorus, are made by other fusion processes (usually tossing in a hydrogen to make up the odd weights). Thus, all the atoms in you, Reader, were made inside giant stars, except the hydrogen in all that water, most of which came from the Big Bang. It may be weird to think that, but you had to come from somewhere.

The famous periodic table of elements, first constructed by the nineteenth-century Russian chemist Dmitri Mendeleev, lists all the elements known to us; many elements are much heavier than iron, but they all exist in small quantities. Most of these heavier elements are indeed harder to make and thus are made in small "trace" amounts (relative to other elements) by a process called neutron capture. In slow neutron capture, which occurs inside stars, first iron nuclei capture neutrons left over from other fusion reactions. The heavy iron isotopes are unstable and typically kick out an electron to convert one of its neutrons into a proton, and thus make the next heaviest element in the periodic table. These in turn absorb more neutrons, and so on, progressively and slowly making heavier and heavier elements. Another form of neutron capture, called rapid neutron capture, occurs during the cataclysmic death of giant stars.

In another 5 billion years, when our own small star dies, it will use up the hydrogen in its hot core, the fusion reaction will peter out and no longer keep the Sun inflated, and the collapse that started 5 billion years ago will recommence. However, given that the Sun will still be very hot, further collapse will heat it up even more, until it reaches 100 million degrees Celsius, when it will start to fuse helium into carbon and then some oxygen, just as the giant stars do. At that point it will expand outward as a red giant (not red supergiant), blowing off much of the remaining gases not in its core, and consume most of the inner solar system (including Earth). The helium fusion or alpha-process reactions in the core will burn through their fuel quickly and cease, but any further collapse will not induce high enough temperatures to trigger new fusion reactions that make heavier elements. And thus at that point, our star will really be dead, and after shedding its remaining hydrogen and helium atmosphere, it will leave behind a very slowly cooling, glowing body made of extremely dense carbon and oxygen, called a white dwarf, with a size about one-hundredth that of our Sun.

The death of giant stars is more cataclysmic but also more productive. Once a giant star has used up its fusion fuel, it also starts to collapse again. However, it is so large the collapse is very sudden and violent, and the rebound of its outer layer off the denser core creates a massive shock wave and explosion, a supernova. It is during the supernova that rapid neutron capture occurs, where the absorption of neutrons by atoms to make elements heavier than iron happens very abruptly. However, the supernova is also important because it scatters many of the products of the star's hard work—that is, the elements it made in its multilayered fusion factory—outward into the galaxy. These elements then load the next generation of clouds with heavier dust that will make new solar systems, but this time with material to make planets. Finally, as noted earlier, supernovae can be important for initiating the collapse of pre-solar clouds; this likely happened in our solar system, as evident, for example, in some meteorite dust grains that contain some heavy isotopes of iron, which can have been formed only during a previous supernova.

Most of the mass of a giant star is ejected during its supernova, but the fraction that remains collapses into a very dense mass. If this remnant is about two to three times the mass of our Sun, the electron clouds holding up the volume of every atom will not be able to withstand the remnant's massive interior pressures, and they will be squeezed out of their orbits into the nuclei of their atom, neutralizing every proton and turning it into a neutron. An atom is about 10^{-10} meters across (called an Ångstrom, or Å), while a nucleus is about 10^{-15} meters across; this size difference is comparable to that between a sports stadium and an ant. The radius of each atom will thus collapse to 10^{-5} of its original value and, since volume goes as radius cubed, the density (mass divided by volume) will increase by a factor of 10^{15}, that is, by a quadrillion. The remaining superdense body is called a neutron star. The density of the neutron star material is so huge that an eyedropper full of it would have roughly the same mass as that of the entire human race.

If the remaining mass of the neutron star is heavier than about three solar masses, then the neutrons squeezed against each other are not able to withstand the pressure, and they collapse into a denser mass, presumably made of quarks, and referred to as quark stars, although these have yet to be unequivocally observed.

If the remnant mass exceeds about five solar masses, then even the quarks cannot resist further compression and the mass collapses into a minuscule volume, which forms the kernel of a black hole. In particular, once this giant mass becomes dense enough, its gravitational attraction is so large that at a finite distance from its center even light cannot escape its pull; the distance at which light is captured is called the event horizon. There is evidence that black holes do exist, including super-massive ones at the center of galaxies.

The final exotic death throes of supergiant stars are crucial for the rest of our story because, after these stars create the building blocks for planets and life, they spew them out into the galaxy with supernovae. To make planets (and you), these big stars have to produce elements heavier than hydrogen and helium in large enough quantities, and then give them up to make new dusty clouds, from which our solar system formed some 5 billion years ago. This requires plenty of giant stars to form, churn out elements, and blow up, and for them to do this often enough to populate the galaxy with planet-making clouds.

Discoveries of many solar systems loaded with planets even in our very nearby galactic neighborhood imply that such solar systems are not rare and thus there is enough material to form them. Thus, if red supergiant stars lived as long as our small star, then most would still exist today and very few would have given up their elements to make solar systems elsewhere. But because of their extreme temperatures and pressures, the red supergiants (and the first stars that ever formed, when the Universe was only a few hundred million years old) rapidly churn out heavier elements and burn through their fuel quickly. This entire process of burning through hydrogen to produce elements up to iron is so fast that a giant star finishes its work, uses up its fuel, and explodes to seed other pre-solar clouds in just a few million years. Thus, giant star formation and death can happen many times and produce enough dusty clouds in a few billion years to allow a solar system to form, as ours did 5 billion years ago. Indeed, major star formation probably peaked about 10 billion years ago, and thus we are possibly even late to the game.

3   SOLAR SYSTEM AND PLANETS

Our solar system and our planet Earth formed almost 5 billion years ago, which was already 9 billion years after the birth of the Universe. The story of how we came to know the age of the solar system and Earth has as much color and controversy as that of the age of the Universe. Scientific thought about the age of the Earth also conflicted with religious doctrines. Yet one of the most famous and contentious debates was not between scientists and religious scholars but among scientists themselves.

In the 1800s, the British physicist William Thomson, aka Lord Kelvin (whom we last saw in chapter 1), calculated that if Earth froze from an initially hot molten state and thereafter cooled like a uniformly hot solid sphere suddenly exposed to a cold surface (of space or atmosphere or ocean or what have you), it would have taken about 20 million years to reach the current state at which it releases heat. Kelvin's Earth was "young" because a solid sphere of mostly rock doesn't cool easily, so to account for the Earth's high rate of heat loss now, it must have been exposed to the cold surface and started cooling fairly recently (geologically speaking). Kelvin validated his estimate by also calculating the age of the Sun. As he understood it, the Sun heats up only as it collapses under its own gravitational self-attraction (which indeed it originally did before hydrogen fusion kicked in), and given its present size and luminosity, it would also be about 20 million years old. That age was certainly much older than Bishop Ussher's biblically inferred 6,000-year-old Earth, but geologists and evolutionary biologists believed this was still not long enough.

Geologists (including Charles Darwin) estimated that it would take at least hundreds of millions of years to build the sediment layers evident in mountains and canyons, especially given the slow rate of deposition by rivers and floods. Biologists had similar estimates for Earth's age in order to account for all the biological

diversity and a rich fossil record accumulated at the snail's pace of biological muta-
tions. But Lord Kelvin—who was an intimidating intellectual figure—considered
such arguments merely qualitative, unable to withstand the rigor of his dual physical calculations. The debate between the physicists and geologists raged on, often
abusively, for decades. In the end, however, it turned out they were all wrong.

The discovery of nuclear decay of radioactive elements eventually settled the debate about the age of the Earth. Radioactivity was discovered in the late nineteenth
century by Henri Becquerel and Marie and Pierre Curie, for which they all shared a
Nobel Prize. Their work showed that large atoms of certain elements, like uranium,
are unstable and spontaneously transform into a different element by ejecting particles from their nuclei. Since many radioactive elements exist naturally in rocks, it
was inferred that the Earth's interior is peppered with radioactivity. The heat released from the energetic ejection of particles during radioactive decay could possibly keep Earth hot enough to reach its present state, even if it froze from a molten
state billions of years ago. This argument, however, originally made by Ernest
Rutherford, does not hold as much water today, since the Earth probably has a
lower concentration of radioactive elements than originally thought; moreover,
radioactivity would have had little effect in Kelvin's static-Earth model. It was also
proposed (by John Perry and Osmond Fischer during the years of this debate) that
fluid convection in the Earth's interior, whereby hot material rises and cold material sinks (as discussed in chapter 4), would invalidate Kelvin's model. In particular, convection could sustain Earth's high heat loss for billions of years by delivering deep, hot material to the surface, while Kelvin's static-Earth model only allowed
conductive cooling of material closer to the surface—and starting "recently"—to
account for Earth's heat loss. Moreover, the discovery of thermonuclear fusion in
the 1920s and 1930s eventually led to the realization that the Sun is not powered by
its gravitational collapse but by hydrogen fusion (as discussed in chapter 2), which
can keep the Sun burning for billions of years.

However, this debate was only truly put to rest with the accurate measurement
of the age of the Earth and solar system in the early to mid-1900s, inferred from
"radiometric dating" of rocks and meteorites. As a radioactive element decays, it
eventually converts atoms from the original radioactive "parent" element (like uranium) to a stable "daughter" element (like lead). The relative amounts of parent
and daughter atoms in a sample can be used to determine mineral age; for example, the more daughter atoms there are relative to parent atoms, the older the sam-

ple. These relative abundances of parent and daughter elements are used along with the decay rate, or radioactive "half-life," to give a reasonably accurate age. This method has solidly settled the age of the Earth and solar system at about 4.6 billion years, although no rocks of that age exist on Earth; the oldest rocks exist in meteorites, most of which are pieces of rock that have fallen to our planet from the Asteroid Belt.

Our solar system formed from the collapse of a giant dusty cloud nearly 5 billion years ago. The collapse of that cloud was probably triggered by the shock wave from a supernova, evidence of which exists in tiny diamonds in meteorites that are embedded with heavy isotopes of iron that could only have been formed by a supernova. The typical initial size of these clouds at which collapse starts is about 1 to 3 light years across—many times the size of our own solar system—just to create a star like the Sun, and probably tens of light years across to create much bigger stars. These sizes are still a minuscule fraction of the extent of our galaxy, which is about 100,000 light years across. Only a fraction of such a cloud eventually winds up in a solar system like ours, and that fraction is called a cloud core. With the successful collapse leading to our solar system, most of the mass of this cloud core fell into the center to make the Sun (as described in the previous chapter). Only a tiny amount of the remaining mass, about 0.1 percent of the Sun's mass, now makes up the solar system's planets.

All the major planets orbit the Sun in a disk, called the ecliptic plane. The formation of a disk-like solar system is caused by the very slow initial spin of the collapsing cloud. As the cloud collapses, the rate of its spin increases, much as twirling ice skaters spin faster when they draw in their outstretched hands. As the cloud rotates faster and faster, the centrifugal effect—which acts to thrust mass outward from the spin axis—resists collapse perpendicular to the spin axis. The same is not true for collapse parallel to the spin axis, thus the cloud can more freely collapse in that direction. The result is that the cloud collapses inward while at the same time flattening like a pancake. The tiny amount of mass left in the disk orbiting around the Sun made up the planets of the solar system.

As nice as that "flattening disk" story is, it leads to one of several major paradoxes about the formation of the solar system. If the cloud really acted like an ice skater all the way to the end, then the solar system as a whole would be spinning much faster and the resulting centrifugal forces would not have allowed it to col-lapse to its present "small size." Even if the initial cloud core was barely spinning at all, it would collapse from such huge distances that it would be like an ice skater holding heavy weights and pulling his or her hands in from miles away, not just pulling in empty hands from an arm's length away.

Distant nebulae, like the dust cloud that created our solar system, are observed to have a very slow spin. The energy (specifically the kinetic energy) of that spin-ning motion accounts for typically a few percent of the cloud's total energy, which is mostly its gravitational energy (that is, the energy released when the cloud col-lapses, which is what heats up the gas and triggers hydrogen fusion and star for-mation). Even with this tiny amount of spin energy, if the huge cloud collapsed to the size of our solar system, the Sun would be rotating much faster than it currently does, and the disk of the rest of the solar system would be spinning around much faster than what our planets' orbits currently allow. However, the centrifugal effect of this spin would never have allowed our solar system to collapse to its current size, in effect leading to Jupiter (which is five times as far from the Sun as is Earth) being somewhere outside of Neptune's orbit (which is thirty times as far away as Earth's). Somehow the solar system shed its spin energy or, not quite equivalently, its "angular momentum" in the process of collapsing. This is referred to as the angular momentum paradox in solar system physics, and it has yet to be solved. There are possible solutions involving everything from magnetic fields to turbu-lence stealing the Sun's angular momentum and ejecting it out of the solar system, but not everyone in the solar system physics community has bought just one expla-nation yet. Regardless, the solar system solved its own angular momentum problem (even if we haven't) and the initial pre-solar cloud collapsed to a nice solar-system-sized disk that eventually let Jupiter assume its current orbit. This initial collapse was very fast (on geological time scales) and probably took about 100,000 years.

Because the angular momentum paradox is given that name, I'm compelled (probably against my better judgment, especially since I've avoided it this long) to explain angular momentum. And the concept will come in handy later, for better or worse. Momentum is a quantity measuring a body's motion (and ability to impart motion to other bodies) in terms of both its bulk and speed. A body's linear mo-mentum is quantified as its mass multiplied by its velocity. A car going 100 kilome-ters per hour has more momentum than a motorcycle going 100 kilometers per hour, and the car will have a bigger impact imparting its motion to another object in a collision. Angular momentum of a rotating body (either spinning around in place or orbiting around a point) is similar but is the body's mass multiplied by its rotation rate (measured most commonly in revolutions per minute), and that quantity is in turn multiplied by the square of the system's effective size. By "effective size," I really mean the distance from the rotation axis to where most of the body's mass resides. So a bike wheel, with most of its mass in the rim, weighing 1 kilogram and spinning at one hundred rotations per minute will have more angular momentum than a thin 1-kilogram spindle or axle spinning at the same rate. The different motion these would impart to other bodies is easily visualized by imagining stopping their spinning with your hands.

Because most of the planetary mass is in Jupiter, and because it is pretty far from the Sun, most of the solar system's angular momentum resides with Jupiter's orbit. However, if the angular momentum of the initial cloud had not been ejected at some time, the Sun would be spinning much faster, and Jupiter's angular momentum would be thousands of times larger than it is now, in which case the planet would be much further out in the solar system than it is today.

The original pre-solar disk was filled with dirty gas, composed of mostly hydrogen, some helium, and various other components of dust and ices from elements made in giant stars over the prior few billion years. All of the parts of the disk were orbiting around the central mass of the cloud, which was en route to becoming the Sun, and that orbiting motion provided the centrifugal effect that would keep the gas disk from collapsing inward further. However, the gas would not quite orbit the central mass as a planet does now.

A planet now orbits the Sun in an exact balance between the inward pull of the Sun's gravity and the outward centrifugal push of its orbiting motion, resulting in a Keplerian orbit (again, named for Johannes Kepler, who empirically inferred the laws of planetary motion in the seventeenth century). But near the center of the pre-solar cloud, the disk was thicker and gas there would be heated up by the forming proto-Sun, which would cause the disk gas pressure to be larger near the center than in its colder, thinner outer parts. That difference in pressure would impart an outward push, from high to low pressure, on the gas that would counter gravity slightly; as a result, the gas would feel a weaker inward pull toward the proto-Sun than would a planet orbiting in a vacuum, causing the gas's orbit to be slightly slower than a planet's, that is, slower than a Keplerian orbit. This effect admittedly sounds pretty esoteric, but it turns out to be the setting for yet another mystery of solar system formation.

The solar system's planets formed from small particles in the dusty gas disk at the same time that most of the disk's mass was falling inward to form the Sun. It would take only several to tens of millions of years for most of the disk's mass to be consumed by the Sun, before it would light up and start fusion. And just before fusion started, the proto-Sun would kill off any further planet formation (as we'll discuss shortly). Thus, the planets, especially the giant ones, had to hurry up and form before this happened, racing against solar ignition, and in the meantime facing various hurdles of their own.

As the dirty gas disk formed, solid motes of dust and ices would stick to each other just by electrostatic forces (like static electricity and other effects, such as the Van der Waals force, which I'll leave for the reader to look up), since they were not heavy enough to attract each other by gravity. Turbulent eddies in the gas probably helped these motes to hover near and twirl around each other long enough to stick together. All of this is not so dissimilar to how dust bunnies accumulate in your home (well, at least my home).

But even to make a small planet, the earliest dust motes (either mineral or icy) would need to grow, and would need to get big enough to start gravitationally attracting more mass to get even bigger. This is easier said than done. When the accumulating motes were tiny (about a micron, on the scale of a bacterium), they would easily float around in the gas disk and move with the gas, all the while sticking to each other by electrostatic forces. Once they grew big enough, say a centimeter or bigger, they would start feeling more the tug of gravity toward the proto-Sun, and less the outward push of gas pressure, and would start orbiting the proto-Sun more like an orbiting planet, in a more Keplerian orbit. Such orbiting chunks would move faster than the gas in the disk and so would experience a headwind and gas drag, which would slow down these chunks and make them spiral in toward the center of the cloud.

If these chunks could somehow get big enough to make planetesimal-sized bodies, like a small asteroid, say 10 meters to 1 kilometer or so across, then they could plow through the gas and feel little headwind and not spiral inward, or the spiral would be slow enough to outsurvive the gas (which is about to be removed, in a few paragraphs). Moreover, at the size of a kilometer or more, the bodies would be heavy enough to gravitationally attract more mass and grow faster yet.

However, for intermediate-sized chunks, between a few centimeters to about a meter in size, the headwind would be very large and would cause the chunks to spiral in rapidly and be lost into the proto-Sun within a few hundred years, which is the blink of an eye in this business. To add insult to injury, these intermediate-sized bodies would be neither sticky nor heavy enough to attract each other and grow, but would instead tend to bounce off one another.

Because all planets were built from tiny grains of dust, they had to accrete mass and grow through this several-centimeters to meter size, but somehow do it quickly enough (despite not glomming together very well at that size) before spiraling in and flushing down the solar drain. In short, they would, somehow, have to grow extremely rapidly past this 1-meter size, in less than a few hundred years, or else they would be lost. This paradox, called the 1-meter hurdle, has also not been entirely solved. However, there is recent progress showing that the push of growing chunks against the gas causes them to cluster; thus, they gather loosely into effectively larger masses while shielding each other from the gas headwind, much like cyclists in the Tour de France.

While the first dust bunnies were accumulating, the collapsing mass at the center was heating up on its way to becoming a star, but even before fusion started, it was getting hot enough to heat up the inner part of the gas disk. The dust motes glomming together in the hotter inner disk were mostly mineral compounds that were not easily evaporated and would eventually make rocks. The outer solar system was cool enough to condense ices and liquids like water, methane, and ammonium. The boundary between these regions is called the Snow Line, which happened to occur not too far from Jupiter's current orbit around the Sun (in a region between Mars's and Jupiter's current orbits).

Because of the gas drag on small particles described above, small icy flakes and chunks falling or spiraling in toward the center of the cloud would evaporate at the Snow Line, and the release of gas would cause a relatively high-pressure zone there. Gas in the disk just outside the high-pressure zone would feel an outward push that would further offset gravity and cause the gas there to orbit the proto-Sun even more slowly, inducing even more headwind and drag on the faster-moving solid particles, which would make them spiral into the Snow Line faster. Gas just inside the Snow Line and its high-pressure zone would feel an extra inward push toward the proto-Sun that would augment the pull of gravity, making the gas orbit faster than the solid particles, giving the particles a tailwind that would hoist them back to higher orbits and spiral outward. In essence, particles would be drawn from both sides toward the high-pressure Snow Line, which thus acted as a trap for icy particles. (Admittedly, this effect is counterintuitive since fluids usually get drawn into pressure lows like a drain, but the interaction of gas and particles in an orbiting disk is more complicated than flow in a bathtub.)

The resulting accumulation of gas and ices at the Snow Line possibly created a fertile ground for forming a giant planet, hence Jupiter. As far as planetary mass (and orbital energy or angular momentum) is concerned, Jupiter is the 800-pound gorilla of our solar system. Other than the fact that we live on Earth, basically most of the important stuff, that is, the mass, energy, and angular momentum of the solar system, is accounted for just with the Sun and Jupiter. But it just goes to show that size isn't everything (or so we Earthlings would say).

Once Jupiter started to form, it accelerated the growth of neighboring giants like Saturn. In particular, Jupiter's gravitational tug would tend to accelerate more slowly orbiting masses outside its own orbit and make them spiral outward. Meanwhile, dust and ice spiraling inward from even higher orbits would merge with this outwardly migrating stream, leading to an accumulation of mass and a feeder zone for another giant planet, like Saturn.

The growth of the first proto-planets from motes of dust was, by necessity, very fast. While planet formation had to overcome various obstacles, like shedding angular momentum and jumping over the 1-meter hurdle, it was also in a race against the Sun. As dust was accreting into chunks, the growing proto-star was eating up disk mass and about to start its hydrogen fusion and light up. Just before ignition, the proto-star heated the inner solar system and ejected gas, in an intense solar wind, that would blow out any remaining dust and gas in the disk cloud that hadn't been captured by sufficiently massive bodies. The final loss of the gas disk and the intense solar wind happened within tens of million years or less after the pre-solar cloud started to collapse into a disk, which is a very short time by geological and cosmological clocks. This means the first proto-planets, especially the giant ones with their massive gas envelopes, had to hurry up and form before all their ingredients were eaten or blown away. Getting these bodies to grow from dust motes to planetesimals and then planets in that short period was a tough nut to crack, and while the solar system figured out how to do it, scientists haven't entirely yet, and so it's another of the many open nagging questions about how to make a solar system.

The terrestrial or rocky proto-planets that did manage to survive and form in the hotter inner solar system were probably at first the size of larger asteroids. Some of these bodies were big enough to heat up and melt; much of this heat was from collisions and some from intense heating from short-lived radioactive elements like unstable isotopes of aluminum and potassium. Once rock melts and starts to refreeze, iron becomes increasingly concentrated in the remaining magma (melted rock) since it dissolves more easily in melt. Eventually, the last melt to solidify is so iron-rich it is heavier than the surrounding rock and sinks to the center of these bodies (if they were big enough to have significant gravity) to form iron cores, and so large asteroids like Ceres and Vesta probably have metallic cores. (Meteorites that reach Earth with pure iron in them—called, logically, iron and stony iron meteorites—are thought to be remnants of these cores that were released after the asteroids were shattered in collisions.) Many asteroids were too small for this to happen and so stayed unmelted and more or less in the same composition as when they first formed; most are called chondrites and are thought to represent the building blocks of our solar system (and many meteorites reaching Earth come from these bodies).

These early planetesimals zipped around the solar system in various elliptical and odd orbits until they eventually collided and only ones in more circular orbits survived; bodies in the same circular orbit, or close to it, would move more slowly relative to each other and thus would collide gently and stick without destroying one another. Eventually, over tens of millions of years, these bodies would make bigger and bigger bodies that would neither get destroyed nor, because of their larger gravity, eject material when they collided violently with other asteroid-sized bodies; and thus they'd get bigger yet, eventually making the terrestrial planets we have now.

Today we have the eight major planets and Pluto, which has something of an identity crisis. Though the International Astronomical Union stripped Pluto of planet rank in 2005, discoveries from the NASA New Horizons mission in 2015 motivated Pluto's repromotion to a dwarf planet. Regardless, we now have drier rocky planets in the inner solar system and the giant gassy and water planets in the outer solar system, with the dividing line between these regions best explained by the Snow Line hypothesis. However, our solar system is not necessarily the norm, and even in our own system, where the planets are now is not necessarily where they formed. The most dramatic examples are Uranus and Neptune, which are far out in the solar system (at twenty and thirty times Earth's distance to the Sun, respectively), and should have had access to huge swaths of disk material to consume and accrete—thus, they should in principle be much larger than they are now. Current thinking is that they formed much closer to Jupiter and Saturn—which would also have been closer to each other—and were thus starved of material by their bigger siblings. Saturn, Uranus, and Neptune were eventually cast outward to much higher orbits, in large part because Jupiter's huge gravitational tug acts like a hammer thrower and tends to eject objects outside of its orbit even farther into the outer solar system. Sacrificing some of its own angular momentum to evict its neighbors, Jupiter would have migrated inward. The shifting around of these huge planets possibly also sent a large number of objects inside of Jupiter's orbit spiraling into the inner solar system, leading to the Late Heavy Bombardment around 4 billion years ago, during which the terrestrial planets were pummeled by impactors. The theory describing our solar system's migrating planets is called the Nice Model, after the research group in the University of Nice in France (not because the model is "nice," although I'm sure it is.)

Finally, while our own solar system seems to have the small rocky planets in the inner solar system, astronomical observations of other solar systems have found Jupiter-sized bodies orbiting their inner solar system, almost as if they were in Mercury's close orbit (they are called "hot Jupiters"); again, the best explanation is that these big planets migrated from where they first formed, as they likely did in our solar system.

But of all the various stories of solar system and planet formation, our own planet presents one of the biggest mysteries of all: how did the Earth get its strange Moon? That we even have a moon as big as ours is freakish, because it's nearly as big as many of the moons around Jupiter and Saturn; the biggest Jovian moon, Ganymede, is only twice as massive as our Moon (a factor of two is nothing in this business and might as well be one). By comparison, Jupiter is three hundred times more massive than Earth and Saturn is one hundred times more massive. So how a dinky planet like ours acquired a big moon is a mystery.

Our unusually large Moon may also have been important for the evolution of life. Lunar ocean tides (that is, the low and high tides) cause tide pools, which were thought by Darwin and others to have been the breeding grounds for early life. Tides also cause intertidal zones, that is, a shoreline that is both wet marine and dry (well, damp) land, and organisms there evolved to live in both environments, eventually triggering the migration (or infestation, depending on your perspective) of life onto land.

The Earth-Moon oddity is about more than just size. The Moon's orbit is about 60 Earth radii away now and it orbits Earth in about a month (really 27 days). However, the Moon started out much closer to Earth, and because the Earth and Moon are hanging onto each other through their mutual gravitational attraction, having the Moon closer made the Earth spin faster—again, as when a spinning ice skater pulls in his or her hands. Indeed, fossils of corals, which have daily and seasonal growing cycles, and sedimentary beds from hundreds of millions of years ago record days that were significantly shorter than they are now. If we plopped the Moon on the Earth, the combined planet would be whipping around with a 4-hour day. Thus, the combined Earth-Moon system has enormous spin to it, much faster than the fastest-spinning planet, which happens to be (again!) Jupiter, with a 10-hour day. The Moon's orbit expanded to its current state because lunar tides on the fast-spinning Earth cause bulges on the Earth's surface that run ahead of the Moon, and the gravitational tug of these bulges pulls the Moon forward, slowly "flinging" (if one can imagine flinging slowly) the Moon to higher orbit. Likewise, the Moon pulls back on the Earth's bulges and slows down our planet's rotation. Thus, while the Earth gives its angular momentum to the Moon, the total angular momentum of the Earth plus Moon remains constant.

Another mystery about the Moon was discovered once man-made satellites and landers could infer something about its interior. Most terrestrial bodies have a rocky outer layer, the mantle and crust, and a pretty big metallic, mostly iron, core, for basically the same reason that the planetesimals have cores of their own, as we discussed previously: heating and melting separates them out. But the Moon's core is tiny, which means that in bulk it has very little iron and is almost entirely just rock. So as far as terrestrial bodies go, it's very weird.

Why did the Earth wind up with such a big and weird moon? This has been a nagging question about our planet's formation for hundreds of years. When I was a kid (in the 1960s) we were taught (unequivocally, I might add) that the Moon was torn from the belly of the Earth and left behind the Pacific basin. This "textbook" explanation, called the Fission Theory, was debunked: it's just too hard to rip moons from the bellies of planets. Go figure.

However, the huge spin of the Earth plus Moon and the excessive rockiness (and lack of iron) of the Moon were major clues leading to what is perhaps the best working hypothesis. Early in the formation of the solar system, when the planets were almost the size they are now but there were plenty of smaller ones zipping around, a Mars-sized impactor—which has been named Theia for some reason (perhaps by the same logic of naming a bomb before dropping it)—is thought to have had a swiping collision with the proto-Earth, not head-on but off-center. This collision would have whacked into the proto-Earth's rocky mantle and stripped a lot of it off, in the process also stripping off the rocky mantle of Theia. Theia's remaining core, having lost much of its momentum, fell into the now melted proto-Earth, which then acquired both metallic cores. Small fractions of both rocky mantles were vaporized in the collision and sprayed into a cloud orbiting the Earth. This cloud eventually (perhaps in a few thousand years) condensed and coalesced into the Moon, which was then of course almost entirely made up of rocky material and little to no iron core. And since the collision was a glancing or sideways blow, it set the proto-Earth spinning rapidly; eventually the Earth transferred its spin, or more specifically its angular momentum, to the Moon's orbit through the tides. This glancing-collision, or giant-impact (as it's called) scenario was first proposed in the mid-1970s by planetary scientist William Hartmann, but it wasn't until the late 1980s through 1990s and the last decade that advanced computer simulations showed that such a collision and aftermath were viable.

Be that as it may, the giant-impact theory and simulations are not without problems, and they haven't answered all the lunar mysteries, such as why the Moon's detailed chemical signatures (such as that measured in the ratios of oxygen isotope concentrations) are so much like the Earth's. For example, if Theia came in fast from another part of the solar system, then why isn't the Moon's chemistry more distinct from Earth's? As in much of science, progress in answering the questions about the Moon's origin has been remarkable, but it is far from complete.

Even though the solar system now has eight major planets and their moons, there is a significant amount of material that was never swept up and used to make planets. Far beyond Neptune's and Pluto's orbits is an enormous spherical cloud enveloping our solar system, filled with small icy bodies; this is called the Oort Cloud (for the twentieth-century Dutch astronomer Jan Oort) and resides about fifty thousand times the Earth's distance to the Sun, almost two thousand times Neptune's orbit distance, almost a light year away. The Oort Cloud is the home of long-period comets, which pass through the inner solar system every 200 or more years; that these comets make such huge, slow orbits and appear to come in from all directions, not just in the plane of the solar system, implies they come in from very far away and from a spherical envelope of icy matter. Closer in is the Kuiper Belt (named for the astronomer Gerard Kuiper—also Dutch and also twentieth century), which is another band of icy cometary material, residing just outside of Neptune's orbit, thirty to fifty times Earth's distance to the Sun. In 2006, Pluto was demoted from a planet to a Kuiper body, especially as more such bodies were discovered (although, as noted earlier, Pluto has been repromoted to dwarf planet). The Kuiper Belt is the home for short-period comets that return in less than a couple of centuries, like Halley's Comet, which reappears in our neighborhood every 76 years. Both the Oort Cloud and the Kuiper Belt retain material that might otherwise have been used for making gassy and water planets and icy moons.

The most notable reservoir of material that could have made terrestrial planets is the Asteroid Belt, which is between Mars and Jupiter. The Asteroid Belt has asteroids ranging from the size of rocks and cars to large, odd-shaped bodies like Vesta, with a diameter of about 500 kilometers, and even larger well-formed spherical dwarf planets like Ceres, with a 950-kilometer diameter (both of which were the subjects of the recent NASA Dawn mission). The entire Asteroid Belt has sufficient material to build a large terrestrial planet, but Jupiter never gave it the chance. The Asteroid Belt is close enough to Jupiter that gravitational tides from the giant planet would disrupt any accreting body once it got large enough. Indeed, Jupiter's tides still affect the Asteroid Belt today, as bodies in the belt that see Jupiter pass by in the same spot every few orbits (called resonances) get pulled out of that orbit, opening up the so-called Kirkwood gaps in the Asteroid Belt. Material ejected from the Kirkwood gaps is thought to be the source for most meteorites that reach Earth.

The Asteroid Belt and all the meteorites that come from it are the best examples of the building blocks of the inner solar system's planets. As mentioned already, certain classes of asteroids (and meteorites) called chondrites have avoided any melting or serious alteration, and even have the essential elemental composition of the solar system retained in the Sun. They are thus thought to be pristine examples of Earth's building blocks. The makeup of these chondrites plays an important role in understanding how the Earth, from its rocky interior to its oceans and atmospheres, was built and evolved (as discussed in the next chapter).

Finally, the inner solar system between Venus and Mars, and thus including Earth, has three populations of asteroids—though these are not as dense as the main Asteroid Belt—called the Amor, Apollo, and Aten asteroids. The latter two have bodies that cross Earth's orbit. Such Earth-crossing asteroids do hit Earth every so often. For example, 65 million years ago, one of these asteroids—perhaps about 10 kilometers across, roughly the size of a small city—struck the Yucatán Peninsula and wiped out the dinosaurs. Asteroid impacts are considered very rare but not impossible events. Although the probability of such an impact is low, the potential damage and fatalities would be so massive that the chance of dying in such an event is not insignificant, about equivalent to that of dying in a plane crash. Therefore, efforts to count and track such asteroids and plan mitigation efforts (most likely by slowly deflecting them, if caught early enough) are serious considerations by government organizations like NASA. While an asteroid impact would be catastrophic for us and many other life forms on this planet, it would also simply mark the Earth sweeping up unused material left from the beginning of the solar system.

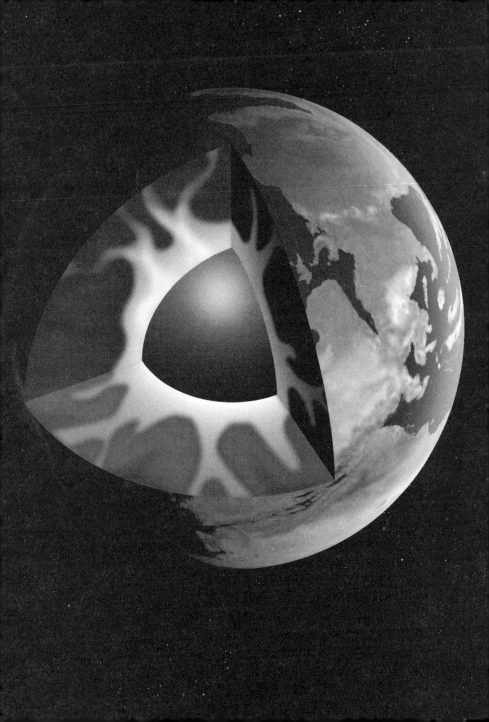

4 CONTINENTS AND EARTH'S INTERIOR

Having created the solar system and planets, we can now zoom in on our home planet and ask how the environment in which we live came into existence. We are, like a host of other organisms, land-based creatures, and so at some point in Earth's history, our very distant ancestors needed some land—namely, continents—to crawl up onto and around. Continents—and in particular our rather special continental crust—are unique to Earth. But to understand where continents came from, we have to travel into Earth's deep interior.

Much of our information about the planets, stars, galaxies, and the Universe comes from astronomical observations, particle physics, spacecraft missions to bodies within our solar system, and meteorites. But to understand anything about the inside of the Earth (let alone the inside of any other planet) means effectively having to "see" through about 6,400 kilometers of rock and metal, to the planet's center. This makes the inside of the Earth more observationally inaccessible than other galaxies, and so it remains one of the great scientific challenges to understand how our own planet works.

Most of what we know about the inside of the Earth comes from seismology, the study of how elastic waves, for example, sound waves, travel through the Earth. But we do not have the luxury of being able to make such seismic measurements on other planets. So far only the Moon has a few active seismometers, left by the Apollo missions, and one of the next missions to Mars (the InSight mission) will finally place a few seismometers there—but that's it, and it's not much. So other observations are necessary. The most basic measurement is just weighing a planet, from which we can find its mass. On Earth we can do this easily enough by putting an object of a known mass on a scale. The weight of the object is the mutual gravitational pull between the Earth's mass and the object's mass; thus, we are measuring not only the object's weight sitting on Earth but also the Earth's weight sitting on the object (so to speak). Knowing also the circumference and radius of the Earth (which was first figured out by the ancient Greek philosopher Eratosthenes), we can also get the mass of our planet, as well as its density, and thus a very rough estimate of its composition. Earth's average density is about 5.5 grams per cubic centimeter (gm/cm^3), which we can compare to water's density of 1 gm/cm^3, or to that of any rock you might pick up from the ground, which is 2 to 3 gm/cm^3, or to that of most metals, which are about 10 gm/cm^3 (iron's density is about 8 gm/cm^3, gold's is about 20 gm/cm^3). So the Earth is denser than most rocks but lighter than most metals, although we also know its substance is compressed to higher than usual densities at the extreme pressures of the deep interior.

The weight of other planets can be found by measuring how a passing or orbiting satellite's motion is affected by the pull of a given planet's gravity; for example, we can also weigh the Earth by knowing the Moon's orbit period (which is just the lunar cycle) and orbit distance (which takes some astronomical measurements, specifically laser ranging these days). We can also get slightly more detailed ideas of a planet's interior layers or structure by how its spin axis twirls around like a wobbling top (a phenomenon called precession). Such information can tell us if the planet has a denser core at its center, and in the Earth's case it certainly does, as do most other terrestrial planets except possibly the Moon, as discussed in the previous chapter. Other satellite measurements give even more detailed data, and rocks expelled by volcanoes give us extra information about the chemistry of some parts of the Earth's interior (more on this below).

But most of the information about Earth's interior comes, again, from seismology. In this case, a very big energetic source of sound, like an explosion, is needed to make big enough waves to go into the interior of the planet and come out the other side. Because of plate tectonics, which we'll get to soon enough, there are frequent big earthquakes that provide this sound source. The resulting sound waves pass through ever-deeper layers with (usually) increasing sound speeds and then arrive at different detectors or seismic stations around the world with different average speeds, depending on how deeply they probed (deeper-probing waves typically go faster). These wave arrivals at different stations are then used to create a picture—or more appropriately take an ultrasound—of the Earth's deepest structure.

Seismology has allowed us to see many layers inside the Earth, but three of them are most distinct and noteworthy: a relatively thin crust of light rock (which in parts has gotten thicker through time as continents have grown; more on that soon), a very thick mantle of heavier rock that takes up about half the radius of the Earth, and an even heavier, mostly iron core, which accounts for the other half of Earth's radius. But because the mantle is wrapped around the core, its volume is much bigger than that of the core. Indeed, the mantle is more than 80 percent of the Earth's entire volume. (This fact is determined by simple geometry: the volume of a sphere is proportional to its radius cubed, so if the core is half the Earth's radius, then it's one-eighth of its volume, leaving about seven-eighths to be taken up by the mantle.)

To measure the density of Earth's layers, seismologists use different kinds of elastic waves that travel through the interior of the planet. The fastest-moving seismic waves are sound waves, caused by the compression and decompression that occur in any medium. The next-fastest waves come from the bending and unbending of material, much like waves on a string, and these can occur only in a solid, since liquids can't unbend on their own if they're bent or sheared. The speed of these two types of waves can be used to infer how easily material is compressed under extreme pressure, and from this its density can be calculated. (There are two other types of seismic waves that are slower and travel only at the Earth's surface; they cause ground shaking and rolling and thus earthquake damage.)

By using these different waves, seismologists have also been able to discern that the Earth's heavy core is primarily liquid, with a density characteristic of a metal like iron. In particular, the Earth's core casts a shadow as the pure "bending" waves radiate out from an earthquake across the planet; since these waves cannot pass through the core, it must be liquid. However, even more detailed measurements have revealed that inside this liquid iron core is a solid inner core, also made of iron, which is most likely where the core is slowly freezing and solidifying (like an upside-down freezing lake). There are indeed even more detailed measurements separating the mantle and even the crust into separate layers, but we can forget about those for the sake of moving on.

Seismology lets us see layering in physical properties like density and even parts of the mantle that might be hotter or colder than other parts. But seismology doesn't give us details about chemical properties. Composition is largely inferred from chemical measurements of both surface and volcanic rocks (which have been erupted from the Earth's interior), meteorites, and even the Sun, which holds the basic compositional signature of the entire solar system. The bulk Earth composition is what we'd find if the layers of the planet were all remixed together into a single homogenous rock, and this is thought to be made of something like the Asteroid Belt's unaltered chondritic meteorites discussed previously, although exactly which kind of chondrite is still debated. Having some idea of this bulk starting composition, and deducing how the mixture separated into different components, which either float or sink depending on their density, gives a reasonable estimate of the composition of Earth's major layers. The core is thus inferred to be primarily iron with some nickel and lighter elements, like sulfur, that dissolve easily in melted iron and were thus carried to the core. The mantle is made of minerals, mostly of magnesium, iron, silicon, and oxygen, which you might recall were all made during the fusion of helium nuclei inside giant stars (through the alpha processes). The crust is made of minerals, involving even more silicon and oxygen, and with a broader mix of lighter metals including (beyond magnesium and iron) calcium, potassium, aluminum, sodium, and so on. (I won't get into the names of these rocks and minerals because I can barely remember them myself.) The cause for the separation of these components from this bulk mixture is melting, which is yet another story.

It's easy to imagine that, after the Moon-forming impact, the Earth was largely molten; it may very well have been molten before this impact, but in some ways that's a moot point (except insofar as it might have affected the impact process itself). Although Earth's geological processes have erased all evidence of such a molten state, there is evidence it existed on the Moon, which has relics of an early magma ocean, literally an ocean of melted rock. Whether the Earth itself had a magma ocean is still an open question, but given the violent nature of planetary collisions and accretion, a magmatic initial state is a good assumption, and thus gives a reasonable starting point for what comes after.

Since many of the larger planetesimals that collided during Earth's accretion probably already had their own iron cores, there was possibly already significant free iron that, being heavy and inserted as big blobs and drops, sank to the center of the Earth early on, likely forming a proto-core even before the giant Moon-forming impact contributed more iron to the core while melting the Earth (yet again).

The Earth's resulting final magma ocean could have been a sizable fraction of the planet's entire volume. As it cooled and froze, the magma ocean continued to separate out components of the Earth, since different constituents of the melted rock mixture—generically referred to as "melt"—freeze at different temperatures and thus typically sink and segregate as they crystallize. Any excess iron still dissolved in the magma would have stayed in the melt till the end (as during the formation of planetesimals) and eventually rained out when the iron-rich melt became heavy enough, contributing the last dregs of iron to the core. Most of the remaining frozen rock layer would make up the mantle, while light buoyant components would have eventually floated up to make an early thin crust. The magma ocean might also have split in half while freezing, with lighter melts staying near the top, and heavier melts, which were compressed to high density at the bottom of the magma ocean, sinking to the base of the mantle. Evidence of this "basal magma ocean" remains today, as there appear to be seismologically detectable pockets of melt still at the bottom of the mantle.

If the magma ocean existed, it froze very quickly (at least the part that didn't sink to the bottom of the mantle), in several tens to a hundred million years, which is fast by geological standards, and it's at this point that an actual geological record, stored in the remaining rocks, starts. Although the solar system is dated to about 4.6 billion years old, those dates are determined from meteorites, not Earth's rocks. The oldest whole rocks on Earth are only about 4 billion years old, and these presumably remained after the magma ocean finished freezing. (There are tiny minerals called zircons found in a few places on Earth that are a few hundred million years older, but the rocks in which they're embedded are not that old.) But there are very, very few of these rocks remaining, because most of the crust that would have floated up during the magma ocean was eroded away and redigested by subsequent geological processes, and perhaps also obliterated by more asteroid impacts, which persisted till about 4 billion years ago. Therefore, 4 billion years ago starts an actual geological era, called the Archean, involving existing rocks, and which takes up a huge swath of geological time (about 2 billion years' worth of the 4.6-billion-year total). The era prior to the Archean, when the magma ocean probably existed, is called the Hadean, after Hades, the Greek god of the underworld.

After the freezing of the magma ocean, the Earth continued to evolve and cool off to the cold vacuum of space, albeit more slowly. This evolution was, and is, largely governed by the mantle. The Earth's mantle is so huge and sluggish it dictates not only how the whole planet cools off to space, but also how it evolves geologically. The mantle is still extremely hot from its early days (after the magma ocean froze), but it is now almost entirely solid, except in a few small but important spots. It is also being heated by energy released during the decay of unstable radioactive elements, like uranium, thorium, and, early on, an unstable isotope of potassium (which decayed quickly but then gave off a burst of heat; in fact, that potassium decayed to argon, which comprises most of the Earth's argon, one of the atmosphere's significant constituents, today). And recall that the heavier radioactive elements (like uranium and thorium) were made by neutron capture during the evolution of red supergiant stars, slowly inside the stars, but rapidly when they underwent a supernova. Either way, the mantle is hot and cooling off to space; more than half that heat is left over from the Earth's formation and magma ocean state, and the rest from radioactive heating.

The mantle, however, does not cool off like a big, hot static rock, but actually moves very slowly. Mantle rock near the colder surface gets cool and heavy and sinks, and rock at the bottom of the mantle near the hot core is warmer and light and rises. This process of hot stuff rising and cold stuff falling is called thermal convection (or sometimes free convection), and it is pervasive in nature—from the mantle to oceans to atmospheres of planets and stars to your cup of coffee. Convection drives hurricanes, thunderstorms, and ocean currents and causes solar granulation on the Sun. It does require fluidity in that the hot or cold stuff has to be able to move under the force of gravity (which makes the hot stuff light and the cold stuff heavy). Although the mantle is a solid, not a liquid, it acts like a fluid over a very long time, similarly to the way glaciers flow, slowly, unless they're melting and surging or falling apart and calving.

That solids act like fluids sounds a little counterintuitive, but as I said in the preface, I won't go fluffy on the science, and so rather than condescendingly tell you that "it's too complicated," I'll give the explanation a shot with a simple model of matter. (Also note that the use of word *fluid* is frequently mistaken as a synonym for liquid. Strictly speaking, the states of matter are solid, liquid, and gas as well as plasma and so on if you really heat it up. But "fluid" refers to how something flows or deforms, not its state of matter, and other ways of deforming are elastic, plastic, brittle, and so on. Thus, a solid can act like a fluid when glaciers and the mantle deform, while gas and liquid can act like elastic substances when sound waves pass through them.)

Imagine a jar that is about a quarter full of marbles (or ball bearings, if you prefer). If the marbles are all sitting at the bottom in the lowest resting position, they're lined up in nice rows and well packed together, usually with each marble sitting in a low spot or divot between a few marbles below it. This case would be like a solid in that the marbles, like atoms, are in an orderly array and basically not moving (if we let them sit). If we swirl the jar hard enough to dislodge the marbles and they start rolling over each other, then it's like a liquid: the atoms are moving around but are still in contact with each other. If we shake the hell out of the jar, the marbles will start bouncing around inside it and fill the volume of the jar: that is like a gas in that the atoms are moving around, filling the volume, bouncing off the container walls, and rarely hitting each other. But let's return to the resting jar, with the "solid" layer of marbles. If we tilt the jar a little, the marbles may be well lodged enough in their little divots not to move, but continued slow tilting will get some of the marbles to leave their divots and roll into the next divots downhill. Eventually we can get this slow movement of marbles from one divot to the next so that the layer has gradually flowed and adjusted to the tilting, but all the while the layer is still an orderly "solid" most of the time (that is, for long stretches of time in between when one marble at a time rolls out of its divot to the next divot). In the actual solid, the moving atoms have left their position in between the other atoms to move to a new solid position. Rocks in the mantle flow this way under stress (stretching and squeezing) and under the influence of gravity moving around light and heavy stuff. But the mantle flows incredibly slowly; our best (or most popular) analogy is that it flows about as fast as your fingernails grow; you don't want to watch them grow (unless you're really bored), but you know they do.

Okay, maybe all that was about as boring as fingernail growth, but it's important because that slow convection of the Earth's solid mantle governs how the entire Earth works. As we will see later, all that convective motion is the cause of plate tectonics and thus earthquakes, volcanoes, mountain building, and the like. Mantle convection also sets the slow pace at which the entire planet cools to space, since the Earth can't lose heat faster than the mantle cools. Convection is one way fluids get rid of heat—that is, by ingesting cool stuff near the surface and mixing it into the hotter interior (like dropping ice cubes in hot tea), and likewise bringing hot stuff from deep down right to the cold surface where it more quickly loses heat. The mantle cools this way, and thus faster than if it were a big static rock, but because it moves so slowly it still cools very gradually. This means the mantle will churn along and drive plate tectonics for billions of years, and we probably need plate tectonics to sustain a stable climate on Earth and hence life for that long; but we'll discuss this later.

The slow cooling of the mantle also ensures that the core doesn't cool too fast, as is evident in the core's still being mostly molten. As we noted already, seismologists, who use the energy from big earthquakes to take an ultrasound of the Earth's interior, have been able to discern that the core is mostly liquid, although there is a solid inner core, which is most likely where the core is slowly freezing and solidifying. Because the outer core (wrapped around the solidifying portion) is liquid, it can easily flow, and since it's made of iron, it is electrically conducting and thus can carry an electrical current. Fluid motion of the outer core is driven both by convection (caused by the core's cooling) and the Earth's rotation. The motion of this electrical conductor in the presence of a weak external magnetic field (coming in from, say, the Sun's magnetic field) causes electrical currents, much the way an electric generator works (a wire bundle spinning inside a magnetic jacket makes electricity flow in the wires). These electric currents then generate their own magnetic field. All magnetic fields are invariably caused by moving electrical charges— either free electrons flowing down an electrical conductor like a wire, or bound electrons as they orbit the nucleus of their atom (which give permanent magnets, like the ones on your refrigerator, their properties). The electrical currents and associated magnetic field generated in the Earth's core become strong and organized enough to power the Earth's entire magnetic field.

Indeed, the Earth has a remarkably strong magnetic field for such a small planet, much bigger than those of the other terrestrial planets. Earth's field is well structured as a predominantly dipole magnet, much like a bar magnet with "north" and "south" poles. Venus, our putative twin, has no detectable magnetic field of its own. The Moon and Mars have patches of magnetized rocks in their crust, and perhaps had their own fields early in their histories, but not anymore. Mercury has a big iron core and does support a dipole field similar to Earth's, but its field is significantly weaker. The giant gassy and water planets in the outer solar system, however, all have significant magnetic fields, and the largest magnetic field is—big surprise—Jupiter's.

Earth's magnetic field extends out through the upper reaches of our atmosphere and even reaches the Moon (with the help of the solar wind blowing the Earth's field out into a whale-shaped structure with a long tail); it also protects us and our atmosphere, as mentioned in the next chapter, from energetic charged particles in the solar wind and in solar storms. Indeed, the field traps these particles high above the atmosphere in regions called the Van Allen belts, which wrap around the Earth; these belts act like magnetic bottles that, when they get too full of charged particles after solar flares and magnetic storms, often spill their contents into the high atmosphere near the North and South Pole, leading to the Aurora Borealis (in the north) and Aurora Australis (in the south). Our magnetic field does not, as portrayed in some Hollywood movies, protect us from uncharged particles and radiation like microwaves.

Evidence that Earth's magnetic field is generated in the fluid core comes mostly from observations that the field originates from inside the Earth (which was figured out in the early nineteenth century by German mathematician Carl Friedrich Gauss), but moves around much faster than mantle-driven geological processes (and than your fingernails grow). Although the field looks somewhat like that of a common bar magnet (made from the mineral magnetite, which has organized magnetism in its crystal structure), it is not from a permanent magnet since the mantle and core are too hot to allow magnetism to be frozen into minerals and iron. The Earth's magnetic field also drifts measurably on the human time scale (which was first noted in the late seventeenth century by Edmond Halley of comet fame), over decades and centuries, and even reverses abruptly every few hundred thousand years (that is, the north magnetic pole flips to the south). So something big, very mobile, and electrically conducting inside the Earth must be powering it (something Halley also guessed at), and the liquid iron outer core is pretty much the only candidate. But it's been only in the last 20 years that the mechanism of core convection powering our magnetic field, called the geodynamo, was demonstrated to work with computer simulations.

However, many details of the geodynamo remain hotly debated. For example, the source of power for the geodynamo is still not entirely known. The power source could be simple thermal convection, with liquid iron near the top of the outer core, at the core-mantle boundary, becoming cooler and heavier and thus sinking. However, iron is also a very good thermal conductor, which means hot and cold convective thermals can be easily diffused away or erased, and so thermal convection might be too weak to be the main power source.

Alternatively, core convection could be driven by composition or chemical differences in the liquid. In particular, the liquid outer core is inferred to be a mix of mostly iron, with some nickel and a small amount of lighter elements like sulfur. When the melt freezes onto the inner-core boundary, the light elements like sulfur prefer to stay in the liquid (that is, they dissolve more easily in liquid) and make it extra buoyant, causing that liquid to rapidly bob up from the bottom to the top of the outer core, thus driving convective motion that powers the geodynamo. That Venus does not have its own magnetic field could be due to its much higher surface temperatures resulting in a hotter mantle and hotter core that have kept its inner core from freezing; this supports the notion that the dynamo is driven by chemical convection associated with inner-core crystallization. But there are other potential power sources for the geodynamo, and which is dominant is still an active area of investigation.

But let's return to the Earth's surface and our original question about the origin of the crust and continents. The crust of any planet generally forms from the lightest melts that come to the surface and freeze; thus, during the magma ocean stage, some of the lightest material spilled onto the surface and made a thin crust, but probably little of that is left behind. Melt rising right from the mantle (or a magma ocean) to the surface mostly looks like a thin runny lava called basalt, perhaps best represented by Hawaiian lavas. Indeed, Hawaii is a good example of how basalt is made today. The Hawaiian Islands originated (and are still forming) over an unusually hot region of the mantle called a "hot spot," which is itself likely caused by a hot convective upwelling or plume that rises across the mantle, perhaps starting from the bottom of the mantle near the hot iron core. Deep in the mantle, the plume material is originally solid and unmelted, but as it nears the surface it melts partially (about 10 to 20 percent or more of it melts) because it's easier to melt at lower pressures (the release of pressure makes it easier for the atoms to move around), and the stuff that melts gets to the surface as basalt. The Hawaiian plume puts out a lot of this basalt, enough to make huge islands of giant shield volcanoes (wide volcanoes with gradual slopes). Other terrestrial planets also appear to have basalt crusts, perhaps made in a similar way; for example, the giant volcanoes on Mars, like Olympus Mons, look like shield volcanoes also.

But the Earth also produces a huge amount of basaltic crust not from volcanoes, like Hawaii, but along long belts of subsea mountain ranges called mid-ocean ridges, which basically wrap around the Earth like the seams on a baseball. But the seams are shoddy at best because the ocean floor rips apart there, and basalt lava rises up from the mantle to fill the gap, freeze, and make oceanic crust. This process is called seafloor spreading, the first observations of which led to the plate tectonic revolution.

Seafloor spreading was predicted by the American geophysicist Harry Hess in the early 1960s, and its discovery (by Cambridge geophysicists Frederick Vine, Drummond Matthews and, independently, the Canadian Lawrence Morley) came soon after. Seafloor spreading is evident in the basalt lavas at mid-ocean ridges, since they have magnetic minerals that record the direction of the Earth's magnetic field as they freeze and cool (like metal shavings on a piece of paper will outline the field lines of a bar magnet below it). Since the Earth's field is, as we mentioned above, reversing periodically, these reversals are recorded in the basalt as the seafloor spreads outward, much like a ticker tape or tape recording (few people these days remember tape recordings, but thumb drives just don't cut it for this analogy, and neither do compact discs). Thus, parallel to these ridges were clearly measurable magnetic stripes showing when the field was pointing up or down, meaning the seafloor was moving outward while recording these events (and indeed these events could be used like time markers to figure out how fast the seafloor was moving).

The discovery of seafloor spreading is what most geoscientists think of as the dawn of the plate tectonic revolution. The idea that the Earth's surface is mobile had been around since the 1920s and 1930s, with the hypothesis of continental drift, although this was a rather different theory from plate tectonics. Continental drift, proposed by German meteorologist Alfred Wegener, contended that the continents move around like icebergs and plow through the ocean crust (which was shown to be impossible), while plate tectonics says the entire surface is divided into giant puzzle pieces that move relative to each other, and the continents are embedded in these plates and go along for the ride. These puzzle pieces are called the

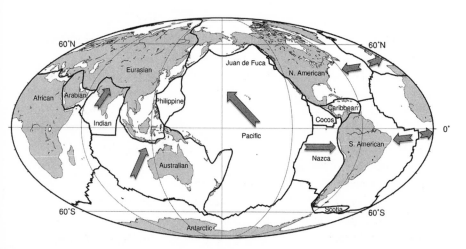

he tectonic plates are like puzzle pieces breaking up the Earth's upper rocky layer, and each plate oves relative to the others. The names of the major plates are displayed, and the arrows on some the bigger plates indicate their motion. The relative motions between plates show the types of ate boundaries, including divergent ones (see the spreading Mid-Atlantic Ridge between the rasian and North American plates), convergent ones (for example, the collision between the dian and Eurasian plates that formed the Himalayas), and transform ones (for example, the San ndreas Fault on the west coast of the United States, between the Pacific and North American ates). Where plates converge at "subduction zones" are also where older, colder plates sink into d cool off the underlying mantle, which is a form of mantle convection. (Figure courtesy of Paul essel, University of Hawai'i at Mānoa.)

tectonic plates, of which there are about twelve major ones, like the Pacific plate (the biggest), and a handful of smaller ones. Many scientists have contributed to fleshing out the modern theory of plate tectonics, but the mathematical model describing plate motions was first proposed independently by Dan McKenzie (from Cambridge) and Jason Morgan (from Princeton). Even so, how and why Earth has plate tectonics (while the other terrestrial planets we know of do not) remains one of our biggest mysteries and an ongoing area of research.

The tectonic plates act like strong solid sheets of colder rock that are up to 100 kilometers thick, but their edges are weak and continuously sliding (on a geologic time scale; on a human time scale the sliding causes earthquakes), thus allowing motion between the plates. As we just noted, in some places these puzzle pieces or plates move apart where the seafloor is spreading. The flip side of these plates moving apart in one area is that they have to move toward each other somewhere else. And indeed they do, in regions called subduction zones. In particular, a plate spreading away from another plate on one of its edges is usually pushing into and diving under a third plate on the opposite side, and that process of one plate diving and sinking below an overriding plate is called subduction. These zones are well delineated at deep ocean trenches, like the super-deep Marianas Trench, where the seafloor is being pulled down by these sinking weights. All this motion is not random and is thought to be the expression of mantle convection as seen at the surface. In particular, a plate subducts because its material cools as it moves away from the hot spreading center where it was created, eventually becoming cold and heavy enough to sink into and cool off the slowly flowing mantle; thus, subduction is equivalent to a cold, heavy convective downwelling.

Geophysicists (like myself) think subduction is not only the manifestation of mantle convection but also the primary driver of plate tectonics. As the cold, sinking part of a plate, called a subducting slab, convectively cools off the mantle, it also pulls the trailing part of the plate at the surface behind it. The main observation supporting this is that plates with significant subduction zones on their edges are the fastest plates; there are a number of plates that have little to no subduction zones and are moving much more slowly—they are probably just being jostled around by the plates that are sinking. The biggest plate, again, is the Pacific plate, which has most of the world's subduction zones and is fast moving, at about 10 centimeters per year.

Subduction zones are also where the most violent and destructive earthquakes and volcanoes occur. Earthquakes do happen on mid-ocean ridges, but they are tiny. Ridges also produce most of the world's lava, but it is runny and flows easily. Where plates neither spread apart nor collide but instead slide by each other, as at the San Andreas and Anatolian faults, the earthquakes are large but not huge, and there is little volcanism since such motion doesn't involve bringing up hot mantle rock to the surface. Subducting plates, however, catch the edge of the overriding plate and pull it down and bend it like a crossbow. The frictional contact between these plates fails or breaks when it can no longer withstand the mounting stress in the bent "bow," at which point the overriding plate snaps back up like the triggered bow, releasing giant earthquakes and often tsunami.

Subduction zones also have big and violently erupting volcanoes, even though they are where cold plates are sinking. So what causes melted rock to come up and make volcanoes in locations where colder rocks are sinking? Volcanic production here is also the key to where most of the continental crust comes from. Indeed, no other planet (that we know of) has plate tectonics, and no other planet has continental crust like ours.

Melting at subduction zones is a little more complicated than melting at mid-ocean ridges or at hot spots like Hawaii. However, in neither case is melting caused because rock is made hotter (the way you normally think of melting something like ice or wax by heating it). At mid-ocean ridges and hot spots, mantle rocks melt because they rise to lower pressure and the release of pressure makes it easier to melt. For subduction zones, melting is made easier by water. Tectonic plates entering a subduction zone have typically been sitting underwater for tens to hundreds of millions of years. When lava comes out at mid-ocean ridges, it starts to react with water and make hydrous minerals (meaning rocks with water or hydrogen in their chemical composition), such as amphibole and serpentine. Sediments washed off continents (which admittedly haven't been made yet) and falling to the bottom of the ocean also get loaded with water (and carbon, which we'll discuss later). By the time a plate gets to a subduction zone, a significant amount of its thin crust has watery minerals, and a lot of these get sucked down into the subduction zone with the rest of the plate (although a lot of sediments are also scraped off and left behind on the surface to pile up). Once these minerals reach a certain depth in the mantle (about 100 kilometers), the temperature and pressure is too much for them to stay hydrated and they release their water—in essence their water is cooked out. This water then leaks off the top of the subducting plate or slab and into the hotter adjacent mantle rock, which itself then gets hydrated. The hydrated mantle rock melts more easily than dry rock (hydrogen weakens the mineral bonds), and so even at the "modest" temperatures next to the cold sinking slab, the moistened mantle is hot enough to melt. This is therefore not a particularly hot mantle melt, but still it rises toward the surface. This melt on its own is like runny basalt lava, although cooler than Hawaiian lavas. Thus, when it hits crust near the surface, it will melt the parts of this crust that can be most easily melted, that is, that can be melted by a "cool wet" melt. Such easily melted rocks tend to be richer in silica (silicon-oxygen molecules or silicates), and these get melted and separated from the rest of the crust. The most silica-rich magma is granite, which is a typical product of such "cool" melting.

The first subduction melting on early Earth could make only a little bit of granite from melting the thin oceanic crust that existed. Even such melting of present-day ocean crust doesn't make much granite (or rocks close to granite) and the resulting island arc volcanoes near ocean trenches (like in the Caribbean and Aleutian islands) can have a lot of the original basalt magma that leaked from the mantle. (The term *arc* is used because subduction zones are shaped like the segments of a circle.) But more granitic rock was steadily produced by continuous melting and remelting of crust, and since granite is light and won't sink in the mantle, it accumulated near subduction zones like floating toys over a bathtub drain. Granitic rock would thus gradually gather into piles of thicker and thicker crust, which is in the end continental crust. Moreover, subduction beneath continents sends wet mantle melts into the thicker crust and causes more melting and separation of silica-rich rocks, making more granite yet. While these silica-rich magmas might be melted more easily, they are very thick and pasty (though less dense), and so they are hard to move; they also hang onto their gas bubbles (made mostly of the water that got the mantle to melt in the first place), which exsolve from the magma as it rises to lower pressures (much like what happens when you take the cap off a soda bottle). Thus, volcanoes made of such magmas, typically the continental arc volcanoes, tend to be taller and steeper (since the thicker, pastier magma can pile up more before spreading) and also they build up much more gas pressure before erupting, which leads to much more violent volcanic eruptions. But eruptions or not, the subduction zone "wet-melting" process is what made continents on Earth.

In total it took about 2 billion years to accumulate the Earth's continents, by slowly and repeatedly melting and separating out silica and granitic minerals from the mantle. But even as continents have sometimes gathered up into giant rafts of thick crust called supercontinents, they have also periodically been broken up and dispersed by plate tectonic motions into normal continent-sized fragments, only to slowly come back together again, hundreds of millions of years later. This cycle of supercontinent accumulation and breakup is called the Wilson Cycle (after Canadian geologist J. Tuzo Wilson). The last supercontinent was Pangea, which started to break up about 200 million years ago, and that breakup is primarily seen as the opening of the Atlantic Ocean along the mid-Atlantic Rise (a spreading center), and this explains why the east coast of the Americas would seem to fit against the west coast of Eurasia and Africa.

To make our continents required two things: plate tectonics and liquid water— lots of water to soak and hydrate the seafloor minerals. Both of these features are unique to Earth, and each probably needed the other to exist. Both plate tectonics and liquid water (as we'll see in a couple of chapters) are probably necessary to stabilize Earth's climate for geologically long periods of time, which in turn keeps surface temperatures clement enough to sustain large quantities of liquid water. Likewise, plate tectonics probably needs water, or at least a cool climate, to persist. Thus, plate tectonics, water, and a temperate climate probably all need each other to exist and are codependent.

How plate tectonics needs water or a cool climate is still an active area of debate. For example, slippery sediments and hydrous melting at subduction zones might keep subduction lubricated and churning along. But Earth's cooler temperatures might also help keep the edges of tectonic plates weak and damaged and slippery. In fact, it might be very hard to have water lubricate all plate boundaries all the way across the thickness of the plates, down to about 100 kilometers, since it would be hard to push water to such high pressures. Something else must keep them weak over most of this depth. Rocks naturally exhumed from such depth near these "rapidly" deforming plate boundaries often show unusual features, such as extremely small sizes of the minerals or grains of which the rock is made (such rocks are called mylonites). These tiny grains possibly help soften the rock and keep the plate boundary slippery, and in turn, the grains become tiny by the grinding and slow damage of rocks at the sliding plate boundary; together these effects lead to a self-softening feedback that would allow plate boundaries to develop and persist. However, mineral grains also tend to grow slowly if left alone (similarly to bubbles growing in foam), which would act to heal and strengthen the rock, and such healing happens faster at higher temperatures. Thus, the Earth's surface is perhaps not only cool enough to sustain liquid oceans but also cool enough to keep these deep damaged plate boundaries from healing. On Venus, with its much hotter surface, healing would be faster and damage weaker and the plate boundaries would be hard to sustain, perhaps explaining why our sister planet appears not to have plate tectonics. However, in the interests of full disclosure, this "damage and healing" hypothesis for the origin of plate tectonics is the author's own scientific research and bias.

If plate tectonics and liquid oceans (and a clement climate) are codependent, it leaves us with a nagging chicken-or-egg question: which came first? This is one of the billion-dollar questions in Earth sciences (it's a big one, so not cheap, yet not as expensive as the Big Bang, which costs trillions of dollars). To answer this we would need to know when (if not how) plate tectonics and liquid oceans first appeared. There are tantalizing but far from definitive clues about the answer.

The last decade has witnessed the discovery of very old tiny zircon minerals (a type of crystal) from 4.4 billion years ago, mostly from only one area of Australia (called Jack Hills). These types of zircons appear to have formed in granites; since most granites are formed eventually by the melting of hydrated rocks, their presence implies that both water and subduction (and thus something like plate tectonics) were present even by that early date. This doesn't explain which came first necessarily—perhaps they occurred simultaneously. Indeed, perhaps if they hadn't occurred simultaneously then neither would have occurred. However, it's also possible, although more rare, to form granites by other means. One could repeatedly remelt rock by, for example, spilling hot Hawaiian-type lava on it over and over again. So the question of which came first, plate tectonics or water, is far from answered at the moment. However, we will explore this question a bit further in the next couple of chapters.

In the process of making the continents on which we live, we've had to see what the inside of the Earth looks like and how it moves, and in doing so we've pointed out another two major oddities of the Earth in addition to its Moon. First, while all the terrestrial planets probably have some form of thermally convecting mantles, only the Earth's version of mantle convection appears as plate tectonics, which (in addition to causing destructive earthquakes and volcanoes) both brings mantle rocks to the surface as magma and drags surface material like water and (as we'll discuss soon) carbon dioxide back into the mantle. The other terrestrial planets have, as far as we know, only a one-way exchange and just spill magma onto the surface through big volcanoes. Second, the Earth has a strong magnetic field, which no other terrestrial planet has, at least not like ours (with the arguable exception of Mercury). This magnetic field extends out through the upper reaches of our atmosphere but is, amazingly, powered by a natural liquid-iron generator at the center of our planet. The deep interior and the vast bulk of the Earth are not necessarily any different from those of Venus, which is almost identical in size to our planet. However, Earth's and Venus's planetary conditions—perhaps their orbital positions relative to the Sun, or perhaps that one experienced a moon-forming impact and the other did not—set them on very different paths, whereby only one of them eventually obtained a magnetic field, plate tectonics, liquid water, and life.

5 OCEANS AND ATMOSPHERE

Life as we know it builds itself from materials mostly found in the thin wispy envelope of gases and liquid water on our planet's surface. We are carbon based and made mostly of water, and we thoroughly rely on plants that transform carbon dioxide and water into sugar. There's more to life than sugar (well, barely, depending on your favorite sugar), and we'll get into those details later. First let's ask, where did this envelope of gas and water come from? For terrestrial planets in our solar system, the fate of their atmospheres (and, if lucky, oceans) was set when the proto-Sun cooked the inner solar system and kept it too hot to condense liquids from the pre-solar gassy disk (back in chapter 3). The outer solar system, beyond the Snow Line, retained a wealth of ices, liquids, and gases, like hydrogen and all the things that can be made with hydrogen (such as water, methane, and ammonia). But the inner solar system was left with a bunch of rocks and no significant atmospheres to start with. Today, however, Venus has a massive atmosphere, Earth's is large enough, and Mars's is thin but significant (although admittedly Mercury has little atmosphere to speak of). Where did these terrestrial atmospheres come from?

There have been two schools of thought, which is a nice way of saying a vigorous debate, about this issue. One theory, called the Late Veneer hypothesis, contends that the Earth and other terrestrial planets had their surfaces violently pummeled by swarms of asteroids spiraling in from the outer solar system during an era called the Late Heavy Bombardment (possibly caused by the outward migration of giant planets, as discussed in chapter 3), around 4 billion years ago, and this would have wiped the surfaces clean of any atmospheres. The ingredients for the existing atmospheres and oceans would have only arrived after this devastating event, delivered from space by cometary material carrying ices of water, carbon dioxide, and other "volatiles" (that is, any substance that evaporates easily) in from the outer solar system. (As discussed in chapter 3, the solar system has two giant reservoirs of comets: in the Kuiper Belt just outside Neptune's orbit and in the Oort Cloud, which is very far outside the main solar system.) Thus, the "veneer" of atmosphere was delivered "late." You get the point.

The other school of thought contends that the atmospheres and oceans were hiding inside the planet, and this has the less sexy name of the Endogenous Origin, meaning the atmosphere came from within the planet. (Thus, Late Veneer could really be called Exogenous Origin.) As we've seen in discussing continental crust, water can be bound up in surface rocks as hydrated minerals, and likewise carbon dioxide can be bound up in rocks called carbonates (a common form of which is limestone or chalk). Most rocks in the mantle can also bind up water and carbon dioxide in various hydrated or carbonated forms, but in pretty tiny quantities; at most these rocks can absorb volatiles in fractions of a percent by weight. However, mantle rocks don't need to take up much water to supply our planet's oceans. The Earth's oceans in total are about 0.03 percent of the mass of just the Earth's mantle (the mass of the atmosphere is trivial in comparison to the ocean), and one could hide our oceans in the mantle several times over without the rocks being particularly wet (maybe just barely damp). Even modestly hydrated and carbonated rocks in asteroids and planetesimals that formed our planet would have buried these components deep inside the Earth as it was growing, and thus the mantle could have retained enough water and carbon dioxide to eventually sweat out an ocean and atmosphere.

But how would water and carbon dioxide ever get out of the planet if they were buried so deep inside? First, if there was indeed a magma ocean, as seems plausible, its crystallization would have released a significant amount of volatile gases like water and carbon dioxide. We can assume that the initial magma ocean contained the volatiles from the original planetary building blocks (such as chondrites). If the entire magma ocean somehow froze all at once, then these volatiles would have stayed dissolved in the final solid mantle, in small concentrations but spread over a large volume. However, the magma ocean would not freeze all at once since it would have been a mixture of several components, some freezing more easily than others. As it froze, the liquid portions that are harder to freeze would retain an increasing amount of water and carbon dioxide, since the liquid can dissolve a lot more of these volatiles than can the crystallizing solids. (A good example of how liquids are better than solids at dissolving most chemicals can be seen in how water is much more efficient than ice at dissolving salt, and even sea ice is pretty close to being salt free.) As the magma ocean finished freezing, the last dregs of melt were very rich in volatiles; some of these melts would have been light and risen toward the surface, while deeper, more compressed melts would have been heavy and sunk to form a basal magma ocean (as discussed in chapter 4). As the buoyant liquids rise to shallower depths and lower pressures, they dissolve volatiles poorly and thus release them (which is why a soda bottle fizzes when you open the cap and release the pressure: the carbon dioxide is suddenly insoluble and makes bubbles). The eventual freezing of these liquids would then liberate most of the remaining volatiles. In total, the last buoyant melts in the freezing magma ocean would at first hoard and then let go of a lot of water and carbon dioxide, releasing them as gas to the surface, probably rather rapidly, geologically speaking.

Although it's likely that magma ocean solidification released most of the early water and carbon dioxide atmosphere, the mantle would have continued to slowly release gases and water even after it solidified. Thus, even if there were no magma ocean, the mantle would still have leaked out the early atmosphere, just more gradually. As we've discussed before, the solid mantle convects slowly, and as hot ascending rock nears the surface and goes to lower pressure it can melt more easily (though only partially, perhaps tens of a percent), and that melt gets dumped on the surface to make some crust, mostly oceanic crust. As noted above, when the melt first forms, it is much better than solid rock at dissolving volatiles, like water and carbon dioxide. Thus, when mantle melts, the water and carbon dioxide dissolved in the rock rush (in a manner of speaking) to the melt, which gets loaded with these volatiles. As the melt rises toward the Earth's surface, it goes to lower pressure and (like the opened bottle of soda) starts releasing its gas; this is what causes volcanoes to erupt, the rapid release of water and carbon dioxide in rising magma. Then as the magma freezes on or near the surface, it does an even worse job of dissolving gas and lets most of the rest of it out. The bottom line is that the small amount of melting of the solid mantle sucks these gases out and delivers them to the surface, and that delivery is entirely by volcanism in all its forms (from violent volcanoes to peaceful deep-sea spreading centers). In the end, between the freezing magma ocean and/or volcanism, it doesn't take much melt to extract volatiles from the giant mantle to make the oceans and atmospheres.

Which hypothesis, Endogenous or Exogenous Origin, is correct? In natural science, it's rarely a clear-cut "either-or" answer, and the best answer is probably that both deliveries of water and other volatiles likely occurred. Which style of delivery was more important is perhaps a better question. One of the primary arguments against the exogenous or Late Veneer hypothesis is that the chemical signature of comets (which can be measured telescopically by the spectrum of light they reflect, or directly in rare instances by spacecraft) is not the same as in the Earth's oceans. The most obvious signature is the ratio of the amount of heavy hydrogen called deuterium (with one proton and one neutron in its nucleus) to normal hydrogen (with just one proton), which tends to be distinctly larger (that is, with more relative deuterium) on comets than on Earth. The range of this ratio for comets, however, is pretty broad and overlaps just slightly with Earth, so the evidence is not entirely a smoking gun. However, other similar ratios, for example, between nitrogen isotopes, show comets and Earth to be very different. Alternatively, the chemical and isotope signatures of meteorites from the asteroid belts, namely, chondrites, easily overlap with those of Earth. Thus, the isotopic evidence generally points toward the ocean and atmosphere not coming belatedly from outer space, but mostly from inside the Earth as delivered during its accretion by chondritic building blocks. Moreover, the Late Veneer hypothesis is predicated on the idea that heavy meteorite bombardment would have wiped out the atmosphere up until about 4 billion years ago (even a little after). But the Australian zircons mentioned earlier suggest that liquid water was present at the surface more than 4 billion years ago despite a hot and hostile environment.

Given all the evidence to date, it appears that the terrestrial atmospheres were mostly degassed or sweated out from the inside of our rocky planets through magma ocean solidification, later volcanic activity, or a combination of both. In this case, the Earth's first atmosphere looked nothing like it does now; if most of it was volcanic gas, it was primarily carbon dioxide and water.

Both carbon dioxide and water are strong greenhouse gases, which means they let in visible light from the Sun, which then warms up the ground; the heat from the ground emanates as infrared radiation and is absorbed and trapped by greenhouse gases, which act like a blanket that warms the surface. With a large amount of carbon dioxide and water, the Earth's atmosphere would have trapped a lot of heat and been extremely hot, with surface temperatures perhaps upward of 200 to 300 degrees Celsius, as opposed to our current cool surface temperature, which is on average about 15 degrees Celsius. Venus is similar in size and composition to Earth, and would have had a similar atmospheric composition, but being a bit closer to the Sun would have had a more powerful greenhouse response and been even hotter yet. In fact, Venus is still close to this state and has surface temperatures close to 500 degrees Celsius. Both Earth and Venus started with very similar amounts of atmospheric carbon dioxide and probably water. Venus today has most of this carbon dioxide still in its massive atmosphere with a surface pressure ninety times that of Earth's surface. (To reach the equivalent of 90 atmospheres of pressure on Earth, we would need to go about 1 kilometer underwater, a depth that can be reached only with submarines.) Earth probably had almost this much carbon dioxide in its atmosphere, at least the equivalent of 60 atmospheres worth of carbon dioxide. However, Earth and Venus ended up in very different conditions today.

Currently Venus has almost no water in its surface or atmosphere, which is still mostly carbon dioxide, and the surface is so hot its rocks glow at night. Earth now has a much thinner atmosphere with very little carbon dioxide, and of course is cool enough to have oceans of liquid water and hence life. If these planets started off so alike, how did they end up so different?

As noted already, the atmospheres of both Earth and Venus would originally have been extremely thick with carbon dioxide and water, and with very high surface pressures and temperatures. On Earth, with slightly less sunlight, its temperature was perhaps just low enough and surface pressure high enough for liquid water to exist at the surface. At our current 1 atmosphere of surface pressure, water boils away at 100 degrees Celsius; but at higher pressure, it boils at a much higher temperature (which is how a pressure cooker works). Specifically, at 60 atmospheres, water would have been liquid on Earth's surface if it were between 200 and 300 degrees Celsius (well, less than 270 degrees, to be exact). This liquid water, in combination with tectonic resurfacing of rocks (as we'll discuss more in the next chapter), may have been enough to start drawing down carbon dioxide by binding it with rocks, which would have gradually cooled off the atmosphere, allowing more liquid water, more drawdown of carbon dioxide, and so on. And the presence of more liquid water and an increasingly cooler surface would also have facilitated plate tectonic activity (as discussed in the previous chapter), and thus kept the drawdown of carbon dioxide going, eventually leaving a small amount of carbon dioxide in our atmosphere and the rest bound up in rocks.

On Venus, with more sunlight, it was likely too hot for liquid water to ever rain out, and what water there was stayed in the atmosphere and kept it extremely hot. Eventually, the Sun's ultraviolet light broke down the water molecules into hydrogen, which would escape to space, and oxygen, being very reactive, would bind with minerals on the surface. In the end only a trace amount of water is left in Venus's atmosphere. Moreover, the lack of liquid water and the high surface temperature probably prohibited regular plate tectonic resurfacing, which otherwise would have helped draw down carbon dioxide. Thus, Earth hit the right combination in which liquid water and plate tectonics could work together to draw down carbon dioxide and hence keep each other going, and in the end allow a habitable surface. Venus never struck this bargain; neither water nor plate tectonics could get a foothold together, and the planet's surface remained hellishly hot, dry, and barren.

Earth's atmosphere today is much thinner than when it started and is now comprised of mostly (almost 80 percent) nitrogen gas and some (almost 20 percent) oxygen, and a smattering of other gases like the remaining carbon dioxide, water cycling in and out of the oceans, and argon, which is inert and stays where it is. Atmospheric oxygen is almost entirely biologically produced and is derived from taking some of the carbon dioxide that was not drawn into rocks and combining it with water to make organic molecules (namely, sugar) and oxygen, in the process of photosynthesis (more on this in chapter 7). The Earth's nitrogen was probably released volcanically as a minor component (relative to water and carbon dioxide) from the mantle, but being relatively inert and unreactive (and at temperatures far above its condensation point), it stayed more or less where it was dumped. Before the massive carbon dioxide atmosphere was drawn down into rocks (and a smaller component into organisms later), nitrogen was a minor atmospheric component, but afterward it was most of what was left.

Although the comparison of Earth and Venus, given their nearly identical size, is most obvious, the evolution of Mars's atmosphere also provides a useful contrast. Mars's current atmosphere is almost entirely made of carbon dioxide and is a hundred times thinner than our own, with an average surface pressure less than 0.01 of our sea-level pressure; this means it would feel like a vacuum to us if we stood on Mars's surface without a space suit. Mars's surface temperature is also extremely cold, with a global average at about −60 degrees Celsius. The Martian polar caps contain mostly water ice (with some carbon dioxide ice), and there is likely a considerable amount of ice in its crustal permafrost regions. Equatorial regions can get warm enough to make ice unstable, but the atmosphere is so thin that the ice mostly sublimates, meaning it directly vaporizes without first becoming liquid. This means Mars's atmosphere does have some tiny amount of water, which eventually snows out over higher latitudes. However, extensive planetary exploration of Mars in the last few decades, much of it searching for signatures of life, has revealed that Mars once had significant amounts of liquid water, evident in riverlike erosion patterns and ancient gullies. Thus, at some point in its past, Mars's atmosphere was thicker and warm enough to sustain liquid water.

There is also sparse evidence that Mars had plate tectonics in its deep past, perhaps coincident with liquid water, and thus perhaps had something like a mutually sustaining water-carbon-tectonic cycle as we have on Earth now, but this is mostly speculation. Regardless, Mars lost its thick atmosphere, leaving it with its current tenuous veneer of gas.

One of the most likely reasons Mars lost its potentially thick atmosphere is that it is just too small to hang onto a warm atmosphere, since the gas molecules could then easily reach high enough speeds to escape Mars's gravity, and thus the atmosphere slowly bled off to space. However, Mars's atmosphere was, at the same time, possibly also being stripped away by solar winds, which even today blow energetic, charged particles (called ions) over the planets, and these particles slowly erode away the atmosphere from its upper layers. Earth's strong magnetic field deflects this solar wind and shields the atmosphere (and us) from these particles and their stripping effect. Venus has no magnetic field and so even now its atmosphere undergoes some stripping; but because its atmosphere is so thick and the planet is heavy enough to hold onto gas molecules, the loss by stripping is slow. Mars perhaps once had a strong magnetic field (inferred from the same satellite observations indicating it had early plate tectonics, which displayed magnetic stripes frozen in the crust, much like on Earth's own seafloor-spreading centers, discussed in the previous chapter), but does not anymore and probably did not for most of its history, and thus its atmosphere was vulnerable to solar wind erosion. That Mars lost (or never had) plate tectonics and a magnetic field is best explained again by its size: it's too small to retain the early primordial heat (left from planetary formation) within its interior, and thus its mantle and core convection have become too weak to power either tectonics or a core dynamo.

The origin of our atmosphere and oceans sets the conditions for the rise of life. However, the structure and movement of both the ocean and atmosphere play important roles in planetary habitability, even aside from having a clement climate (which we'll discuss in the next chapter).

The lowest layer of our atmosphere is called the troposphere, and this is where we think of weather and winds taking place. This layer is on average about 10 kilometers thick (although it's thicker at the equator and thinner at the poles), and is where the atmosphere undergoes thermal convection not unlike the mantle's thermal convection. In this case, the solar heating of the ground warms near-surface air, which rises up, cools off, and then drops back down again, somewhere else than where it started. Thus, the troposphere is warmer at the bottom and colder at its top. As we'll see soon, the convective motion in the troposphere involves some complexity but basically is what drives winds and weather.

Above the troposphere is the stratosphere, whose temperature increases with height. Air at the top of the stratosphere is, therefore, warm, buoyant, and stable— that is, it won't sink—thus aerosols and volcanic dust are easily trapped there. (The stability of the stratosphere also keeps it from convecting and thus it has little turbulence, which is why commercial aircraft fly in the lower stratosphere.) The high stratospheric temperature is due to the presence of ozone, which is a molecule made of three oxygen atoms. Both the production and breakdown of stratospheric ozone (from and to normal oxygen molecules, which have two atoms) involve absorption of certain kinds of ultraviolet radiation coming in from the Sun, which causes the stratosphere to soak up solar energy and heat up. This effect is also very important for shielding life on the surface from damaging ultraviolet radiation. In this sense, the rise of life and production of oxygen were self-reinforcing since the latter created a protective ozone shield. And this process also emphasizes why ozone loss is disastrous, and why the Earth's polar ozone holes, which were theorized in the 1970s and discovered in 1985, are of such concern and have led to international actions and pollution regulation to repair them.

The stratosphere extends up to altitudes of about 50 kilometers, and above that is the even more rarified mesosphere, which extends up to about 100 kilometers and is efficient at radiating away heat and is thus cooler than the stratosphere. Above the mesosphere is the much hotter but much thinner thermosphere (up to about 600 kilometers), and above that is the exosphere (up to at least 10,000 kilometers or more), beyond which is interplanetary space. The upper mesosphere, thermosphere, and lower exosphere have significant concentrations of energetic ionized atoms and are collectively referred to as the ionosphere, which acts as a natural channel for carrying radio waves around the globe.

Let's return to the troposphere. Convection in this layer is driven by solar heating, which is strongest near the equator in the tropics, where sunlight is direct, and weakest at the poles, where sunlight is more diffuse. If the Earth were not spinning, convection would appear as hot air rising up from the heated ground at the equator to the top of the troposphere, then traveling toward the poles, where it would get cold, sink, and return to the equator along the surface. However, the Earth is spinning quite fast, and air sitting on the ground at the equator is moving toward the east (from the perspective of a stationary point hovering over the Earth) at a very high velocity, traveling the circumference of the Earth in one day (that is, 40,000 kilometers in 24 hours, which is 1,700 kilometers per hour). Air sitting on the ground closer to either the North or South poles is moving more slowly, since it travels around smaller circles than the equator in 24 hours; at the poles it's not traveling at all but just very slowly spinning around in place. Thus, air upwelling at the equator has a high eastward speed, and as it rises up and then moves toward the colder poles, it is progressively moving east faster and faster relative to the ground. Thus, while this warm upwelling air attempts to travel to either pole, it is, relative to its surroundings, being deflected more and more to the east until it is essentially moving entirely east along a circle of a given latitude. Eventually it loses its heat and sinks down along this same latitude line, which on Earth is about 30 degrees (north or south—think Florida or Perth, Australia). That cool, sinking air impinges on the ground and spreads out both to the north and south. The ground-hugging air moving toward the equator finds itself being deflected west relative to its surroundings, which are, like the equatorial ground, moving faster toward the east. Those westward-deflected air currents make the trade winds, which are the prevailing winds in the tropics. That entire circulation of warm air rising up from the equator, traveling to plus or minus 30 degrees latitude, then cooling, sinking, and spreading back toward the equator makes up the Hadley cell circulation. Conversely, the ground-hugging air spreading out from the cool, sinking current at plus or minus 30 degrees latitude but traveling toward the poles is (like the original equatorial upwelling) deflected east, and this makes up the mid-latitude westerlies, which are the prevailing winds for most of the continental United States and Europe. (The terms *easterly* and *westerly* can be a little confusing since they describe things that come from the east or west, respectively; thus, the westerlies are winds that blow east.)

Finally, the cold air at the poles, which tries to spread along the surface toward the equator, moves into surroundings that are moving much faster to the east and thus the airflow is deflected west relative to these surroundings. These currents are called polar easterlies, which are the prevailing winds down to plus or minus 60 degrees latitude (for example, over Alaska or Antarctica). The Northern and Southern hemispheres of Earth each have three of these counter-rotating convection cells or rolls that wrap around the Earth parallel to the equator; they are responsible for transporting hot air from equator to pole and cool air from pole to equator, and in the process they drive the prevailing winds of the planet, which are essentially the bottom of each of these cells. These prevailing winds basically drive the direction of all weather patterns (those and the jet streams, which are on the top of and in between each convection cell) and were important for ocean navigation and travel during human expansion between continents.

The strong trade winds also push tropical ocean waters westward, which on hitting the western boundary of ocean basins drive currents north and south, leading to circulation patterns like the Gulf Stream, which brings warm waters to the North Atlantic and gives New England and western Europe their temperate climates. The warm water in the Gulf Stream does eventually cool off in the North Atlantic, where dry, intense westerly winds also drive evaporation and make the water there extra salty. Since the water there is very cold and salty, it gets heavy and sinks vigorously, a process called thermohaline convection. Such ocean currents, driven by both

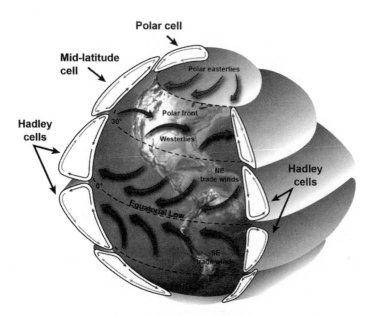

Atmospheric convection acts to transport hot air from the tropics to the poles, and cold air in the reverse direction. Rotation of the Earth, however, breaks the convective circulation into three counter-rotating cells in both the Northern and Southern hemispheres. Air flowing at the bottom of each cell is deflected east or west because of Earth's rotation (and depending on whether the air is flowing toward or away from the equator), and these currents comprise the prevailing winds in the Earth's atmosphere. (Courtesy Barbara Schoeberl, Animated Earth LLC.)

winds and thermohaline convection, are largely responsible for global ocean circulation, mixing, and overturn, which take centuries to complete. This long overturn and mixing time also controls how long it takes for the ocean to adjust to changes in atmospheric temperature and greenhouse gas concentrations (as discussed in the next chapter).

Atmospheric convection cells also determine how water is transported across the planet through the atmosphere. The intense heating at the equator also evaporates a lot of water and this is carried in the warm upwellings. As these upwellings travel to higher altitudes and then spread horizontally north and south, the air cools and the water condenses, creating clouds and rain (hence why the tropics are so humid and rainy). By the time this air reaches its downwelling points at plus or minus 30 degrees latitude, it has lost its water, and so when it descends it is very dry and thus tends to dry out the ground on which it lands; this leads to aridified zones like the Sonoran and Sahara deserts and much of inland Australia as well as Mediterranean environments where arid land meets the sea, such as in (of course) the Mediterranean and much of California. These zones of different climate and humidity played important roles in the development of agriculture and thus the tides of human history and prehistory.

Much of Earth's atmospheric (and thus ocean) circulation is governed by our planet's relatively rapid spinning motion, as described above. Venus has very little spin and even revolves slowly backward (that is, opposite to Earth's spin and that of most other planets in our system) every 243 days, which is even slightly longer than a Venusian year (about 225 Earth days long). Why Venus has such a slow and odd rotation is one of many mysteries about our sister planet. Despite its sluggish spin, Venus has strong winds, which, in its near-equatorial upper atmosphere, flow opposite the direction of planetary rotation (on Earth, the top of the Hadley cell flows in the same direction of rotation). Mars's spin is almost exactly the same as Earth's (for no apparent reason, just chance probably), and even with its very thin, mostly carbon dioxide atmosphere, Mars has something close to a Hadley circulation transporting heat and even water vapor from the equator toward the poles; this circulation also excites vigorous winds that can drive massive dust storms that sometimes engulf the planet for months at a time.

Although I'm being Earth-centric and giving only sparse comparison to our nearby terrestrial planets, it would be too provincial not to mention the fascinating atmospheres of Jupiter and Saturn. Both planets have pretty much the same composition of the pre-solar nebula, which was still basically the same as the Universe after the Big Bang, with some modification. This means they are mostly hydrogen, with less helium and small amounts of heavier elements made in supergiant stars. Despite the immensity of both planets, they spin more than twice as fast as Earth, each with about 10-hour days (Jupiter's spin is a little faster), but both receive much less heat from the Sun than does Earth (Jupiter receives twenty-five times less solar energy and Saturn about a hundred times less). Both giant planets have bands of jet streams and clouds that (in a simplified sense) represent many Hadley-type convection cells; however, probably most of the energy driving this circulation comes from heat loss from inside the planets. The bands of winds, called zonal winds, are extremely fast and on Saturn can reach over 1,600 kilometers per hour (the fastest winds on Earth in a tornado are at best around 500 kilometers per hour). Both planets have huge cyclonic storms, vaguely similar to cyclones on Earth (for example, hurricanes and nor'easters, although both involve water evaporation and condensation as fuel sources) but much, much larger. Saturn has a massive cyclone at its North Pole, and Jupiter's famous great Red Spot is a cyclonic storm that is larger than our entire planet and has persisted for over a century.

While Earth does not have the biggest, hottest, coldest, fastest, or slowest atmosphere, it is unique among all the planets of our solar system for one fascinating reason: its atmosphere is completely different from the way it started. All the other planets have nearly exactly the same atmospheric composition they were dealt more than 4 billion years ago. But because Earth turns itself inside out with plate tectonics, because Earth rained out all its water to make oceans, because Earth developed life, our atmosphere today has no resemblance to its original state. In the end, no known planet has altered or evolved its surface environment as much as has Earth.

6 CLIMATE AND HABITABILITY

Unlike the other planets of our solar system, Earth developed a temperate climate that could sustain liquid water and hence life, or at least life as we know it. The first life that could get a foothold on Earth was microbial, and this occurred a few billion years before we humans would ever consider Earth habitable, let alone hospitable. But even today, we have found microbial life in Earth's most dreadful environments, from above-boiling temperatures to highly acidic volcanic ponds, so the definition of "habitability" has a pretty broad meaning. This implies we might find that microbial life exists or existed at one time on other planets that have conditions no worse than our worst conditions. Liquid water seems key for life, so the list of other possibly habitable planets includes perhaps Mars and some of the icy moons orbiting Jupiter and Saturn that have evidence of liquid water, like Europa and Enceladas, respectively. Regardless, we certainly know our planet developed a particularly stable and clement climate that allowed life enough time to become complex and multicellular.

The planetary conditions necessary for life to exist usually begin with the classical notion of the "continuously habitable zone." This zone is simply the "Goldilocks" orbit in any solar system that is the right distance from the parent star to allow liquid water at the surface. In other words, the planet is not so far away from its sun as to cause all water to freeze (as perhaps on Mars, although this is becoming increasingly suspect) and not so close as to cause all water to vaporize (as on Venus). This notion is still used by astronomers discovering terrestrial planets in other solar systems, since the primary observables they have, at least for now, are the distance from the planet to its host star, and sometimes the planet's mass and/or size.

This orbital habitable zone is also an important component of theories about the probability of finding intelligent extraterrestrial life. By "intelligent," I mean life that can transmit signals from a planet, like radio waves carrying organized information. Whether extraterrestrial life forms would consider our radio waves signs of intelligent life is a matter of debate, but provided we don't misinterpret the Alpha Centauri versions of *Get Smart*, *Star Trek*, or *Bonanza* (a few of my favorites growing up), then the search criteria should be self-evident. The probability of finding these signals was expressed in the famous Drake Equation (after the American astronomer Frank Drake) as the product of various probabilities concerning, for example, the likelihood that stars have planets at all, that at least one of them is in the habitable zone, and the length of time any potential life form has to emit radio waves (in a time interval in which we can detect them, not too soon and not too late). The probability of any given solar system developing life that can emit radio waves in the right period of time to reach us is astronomically small. However, the number of potential stars in our galaxy that could sustain life over a long evolution (generally smaller stars, which burn for billions of years) is in the billions. Thus, if even a tiny fraction of these billions of stars supported radio-transmitting life, there might be millions or at least many thousands of them. In this case, one might expect to see at least one bad extraterrestrial television show cross our radio telescopes, but we haven't, not yet; and this leads to the famous question posed by the physicist Enrico Fermi: Where is everyone? Either the conditions for forming intelligent life are far more complex than originally assumed, or extraterrestrials went straight to cable television.

The conditions for forming life, complex life, and even technologically advanced life are probably more involved than can be characterized by astronomical position and orbit radius. In other words, more than just sunlight determines the conditions for our clement climate. For example, in our solar system the Earth is naturally assumed to be inside the orbital habitable zone (especially given all the overwhelming empirical evidence that it is, indeed, inhabited). However, if Earth had no water vapor or carbon dioxide in its atmosphere, there would be no greenhouse warming and the planet's surface would likely be frozen over, covered in snow and ice, which it very likely was during some periods in the distant past (as we'll discuss below). While some pockets of liquid water might exist beneath the ice cover, Earth would not be receiving much if any solar energy to power life (given the high reflectivity of snow and ice). If life could get energy only from other sources like volcanism, then that would require volcanism in addition to the right orbit. On the other hand, if all of Earth's primordial carbon dioxide—at least 60 atmospheres' worth, now bound up in the crust—were in the atmosphere, greenhouse warming would likely cause an inhospitably hot surface. While, as noted above, there are certainly some microbes that can survive and thrive in extremely hot and cold temperatures, they have not developed beyond their microbial state, at least not on Earth. Thus, even the end-member possibilities for Earth in the orbital habitable zone—either frozen or scorching—would probably at best be habitable only by single-cell microbes. In short, orbit isn't everything. But then what are the conditions for habitability? This also remains another million- or billion-dollar question.

The Rare Earth hypothesis (posited by geologist Peter Ward and astronomer Donald Brownlee) is a nice, albeit controversial, example of a theory that attempts to answer Fermi's question. This hypothesis contends that conditions for habitability on Earth are, as the theory's name implies, an extremely rare combination that has allowed Earth to evolve animal life and hence humans. More specifically, the right combination of favorable conditions is so improbable that the number of extraterrestrials emitting radio waves is too minuscule to be successfully detected in the limited time of our observations. Hence, the answer to Fermi's question is that the galaxy is a lot more like the Gobi Desert than Hong Kong or Paris.

According to the Rare Earth model, our planet satisfied all the normal astronomical conditions by being in the right place in the galaxy, meaning not too close to the galactic center, which is dense with stars and intense radiation emitted by matter falling into a supermassive black hole. Earth also formed at the right time in the Universe to have the building blocks for life. And our planet also sits in the right orbital habitable zone in our solar system, one that permits not just liquid water but the coexistence of water as gas, liquid, and solid ice (which, it turns out, are all important parts of the climate system; more on this below). In addition to the astronomical conditions, Earth has plate tectonics to stabilize climate (more on this to come). It also has a large moon to drive amphibian evolution in tidal zones; that is, organisms in these zones would have had to survive both underwater and in air since the tide moves the shoreline large distances in and out, which then promoted the migration of life onto land. Earth also has the "right" tilt of the planet's spin axis to allow strongly varying seasons that help drive biological diversity. Moreover, the Earth has experienced extinction events caused by Earth-crossing asteroids, large volcanic events (for example, the massive end-Permian extinction event about 250 million years ago might have been due to huge lava flows in Siberia releasing toxic gases and burning out extensive coal seams to drive global warming), and supercontinent formation, which causes loss of coastline and associated marine ecosystems. Each extinction induces an ecological reset that helps drive biological diversity and evolution.

Unfortunately, there is just one Earth that we know of, and thus we don't have much data to show whether a rare combination of many conditions is absolutely necessary for habitability. That is, we don't know whether a few of these conditions are good enough or if you need all of them. In short, we have one data point, because we know of no other terrestrial planet that has plate tectonics, liquid water, or a big moon. In time, this lack of data will change because searches for planets orbiting other stars have already found a number of terrestrial planets. We'll eventually see whether these planets have the various conditions required for life, although it will take much higher astronomical resolution and cleverness to see the details (like oceans and plate tectonics).

We also don't know if some of these rare conditions are independent (making their simultaneity really improbable) or are related (making their simultaneity a given). For example, the presence of liquid water and plate tectonics (and associated processes like volcanism and supercontinent cycles) may very well depend on each other, and so their coincidence may not be . . . well . . . a coincidence. In short, perhaps any terrestrial planet with liquid water will also have plate tectonics, but we just don't know yet. Similarly (as has been argued by critics elsewhere), the Rare Earth notion assumes that these conditions are necessary for animal life as we know it—thus, in a sense, these conditions are just a recipe to get complex life exactly as it is on Earth, not a generalizable theory for complex life elsewhere. Certainly, while it's the only recipe we know, it might not be the only recipe possible to get something else, some other life that we just haven't conceived of yet. In short, we are really just too provincial, living on our one planet without knowing enough about our own solar system and not enough yet to say what other life forms might be like.

Whether or not there are other modes of habitability available, we do know some things about habitability on our planet, and because this is our home for the long haul, it's useful to review and understand these things. And when we refer to habitability, we really mean a stable climate that provides liquid water and a steady supply of building blocks (nutrients) for life, and that doesn't try to wipe us out every few million years or so.

The most important ingredient in our climate is the sunlight we receive. At any given moment, our planet receives about 170 quadrillion (1.7×10^{17}) watts of energy from the Sun. One strong lightbulb gives off about 100 watts, so this would be almost 2 quadrillion lightbulbs illuminating one side of our planet at any moment, or about thirteen such lightbulbs in every 1-square-meter patch of area (most rooms in houses have about a 25-square-meter footprint, typically illuminated by two lightbulbs). Most of this radiation arrives as visible light, which is why, having evolved on this planet, we can see visible light. A fair amount of sunlight is ultraviolet (which is why we wear UV protection in our sunglasses and sunscreen, although a lot of the most damaging UV is absorbed by ozone in the stratosphere, as discussed in the previous chapter); some sunlight is in near infrared, that is, pretty close to just red light.

Some surfaces on the Earth absorb this incoming sunlight and some reflect it back to space. Oceans are dark and absorb a lot of light. Continents are lighter and reflect back some of it. Ice, like that which covers most of Greenland and Antarctica, reflects back essentially all of it. In total, the Earth's surface absorbs about 70 percent of the incoming sunlight (the other 30 percent leads to an effect called Earthshine, which is similar to moonshine, but not the toxic alcoholic version), and so it heats up and radiates energy back out as heat or, equivalently, infrared radiation. If there was no atmosphere, the surface would heat up to an average temperature of only about −20 degrees Celsius, which is very cold and well below freezing (although some parts of the globe would be warmer and some colder). However, our atmosphere has two important gases, water vapor and carbon dioxide, which absorb the reemitted infrared (meaning infrared photons are absorbed by exciting vibrations in these molecules), thus trapping the heat and acting like a blanket. Although neither gas is a major component of our atmosphere (which is, again, mostly nitrogen and oxygen), they are potent greenhouse gases and provide enough of a blanket to let the planet warm to an average temperature of about 15 degrees Celsius.

Our climate is thus very sensitive to how much sunlight our planet absorbs and reflects, and how much greenhouse gas we have in our atmosphere. Therefore, climate stability and habitability greatly depend on whether or not there are big swings in these two effects. Sunlight has in fact been getting steadily stronger since the Sun first started its fusion process, and was about 30 percent weaker in its earliest stages than it is now. But the sunlight we absorb also varies because the ice caps grow and shrink (changing the amount of light reflected), Earth's spin axis (which goes through the North and South poles) twirls and wobbles, Earth's orbit around the Sun fluctuates, and the Sun's output itself oscillates with an 11-year cycle.

Fluctuations in the amount of greenhouse gases are also very important, depending on their potency and the time they spend in the atmosphere. Water vapor is the most important greenhouse gas in terms of its pure blanketing effect. Yet the amount of water in the atmosphere doesn't vary much because the atmosphere is in contact with the oceans and, on average, it is saturated with water vapor and can't hold any more. If the atmosphere is too dry, it will eventually absorb water through evaporation. If it is too wet, it will rain water out. For this reason, air always evolves toward being saturated, neither too wet nor too dry. Therefore, if we suddenly push any more water vapor into the atmosphere, say with a volcanic eruption, most of it will rain out (and given rapid atmospheric circulation, water will rain out well before the extra water vapor can cause any significant greenhouse warming). The water-saturated state of the Earth's atmosphere is also important for ensuring the hydrological cycle of evaporation and precipitation, which we'll soon see is crucial for the planet's overall tectonic thermostat. In contrast, Venus's atmosphere may have been so hot as to always be undersaturated in water, that is, it could take on more water vapor without raining it out. Water vapor added to Venus's atmosphere (either from volcanic degassing or possibly ocean evaporation, if there ever was a Venusian ocean at all) made the atmosphere hotter and thus even more undersaturated, which would cause it to evaporate more water, which would make it hotter yet, and so on, leading to what is termed the runaway greenhouse effect.

Methane is also a very potent greenhouse gas, but it now exists in small (albeit steadily growing) quantities, although it was perhaps much more abundant when life first started and the Sun was dimmer. Methane currently lasts in our atmosphere less than a decade because it effectively reacts with high levels of atmospheric oxygen (more specifically, with oxygen radicals in the stratosphere) to make weaker greenhouse gases, carbon dioxide, and water.

Carbon dioxide is a more potent greenhouse gas than water, but less than methane; however, it has a unique story regarding its residence in various parts of the Earth. There was at one time a massive amount of it in the atmosphere, but it is now mostly stored away in the crust, and to a lesser extent the oceans and biosphere (more on that soon). But when even a small part of that huge hidden reservoir of carbon dioxide is released, it takes a very long time to get it back out of the atmosphere; it doesn't rain out, like water, or react away quickly, like methane. The fastest and most effective sink of carbon dioxide is by dissolution into the ocean, but even this is very slow (as we'll discuss a bit later), thereby causing the gas to linger and accumulate for centuries or longer, thus having a big effect on climate.

There are various important natural feedback mechanisms on Earth that amplify or dampen swings in climate, and several have to do with carbon dioxide. If a feedback loop is positive, it amplifies climate variations; if it's negative, it stabilizes climate. Plate tectonics in particular provides an important negative feedback that stabilizes climate for hundreds of millions of years. Moreover, plate tectonics grinds along regardless of the weather, season, or climate, and so it keeps its negative feedback churning, no matter what happens at the surface. In fact, geophysicists like myself enjoy annoying our climate science colleagues by claiming the most important part of climate science is plate tectonics. It might even be true.

The plate tectonic feedback is called the tectonic or geologic carbon cycle and has a few moving parts. First, plate tectonics delivers fresh minerals from inside the Earth, from the deeper crust and mantle, to the surface. This happens by volcanism at mid-ocean ridges where plates are spreading apart, and also by volcanism and mountain building over subduction and collision zones where plates dive under each other and sink into the mantle, and the continents drawn into these zones get squeezed and folded and pile up. It also happens at oceanic hot spots like Hawaii, but this accounts for much less resurfacing activity. After the minerals are brought to the surface, they chemically react with both water and carbon dioxide, in rainwater as well as in rivers, lakes, and oceans. In particular, carbon dioxide dissolves in water—especially rain droplets because of their high surface exposure—to make a weak acid (actually, carbonic acid, which is the same acid as in carbonated drinks), which chemically reacts with silicate minerals to make carbonate minerals (like in limestone and marble), and in this way the carbon dioxide is drawn out of the atmosphere through water to bind with minerals to be stored in rocks. If these reacted minerals were left in place, they'd make a crusty film of carbonates that would keep the deeper minerals from reacting, and eventually the extraction or "drawdown" of carbon dioxide would stop. However rain, snow, rivers, and glaciers erode away these minerals as they react and wash them into the sea. The erosion helps expose the tectonically delivered fresh minerals that continue to react with carbon dioxide and draw it down.

Erosion by itself would wear down the surface of the Earth to a flat (or, globally, a cue-ball round) surface below the ocean that would shield it from further erosion and slow down or kill off the carbon dioxide extraction (depending on how deep it is in the ocean, but that's a complexity I'm going to avoid). But plate tectonics not only brings up fresh minerals but also continuously builds up volcanoes and squeezes up mountains and keeps the erosion cycle going. As these eroded minerals wash into rivers, lakes, and eventually the ocean, they continue to become carbonated since these waters have a lot of carbon dioxide dissolved also as acid. Today a lot of the ocean carbonation reaction is biologically mediated by shell production in coral reefs and plankton—like foraminifera and coccolithophores—but this reaction would go on regardless. Because of this ongoing carbonation reaction, most of the original carbon dioxide atmosphere of the early Earth, about 60 atmospheres' worth, is bound up as carbonates in the ocean floor and ancient ocean floors that were pushed and uplifted by plate tectonics into mountains and continents. So without this geological drawdown of carbon dioxide, our atmosphere would be much like that of Venus.

However, the geological storage of carbon dioxide isn't truly permanent. In particular, subduction zones (where tectonic plates dive back into the mantle) eventually suck seafloor carbonates into the mantle. Some of the carbon dioxide in these rocks is cooked out at the high mantle temperatures, dissolved readily in mantle melt above the subducting slab (although the melting is due to the water that is also cooked out of the slab, as discussed in chapter 4), and comes back out as gas in volcanoes. However, some of those carbonates survive cooking and perhaps get drawn down and mixed into the deeper mantle. Indeed the mantle is thought to retain a lot of carbon anyway, although not in high concentrations. But given the mantle's immense size, the net amount of mantle carbon is likely to be much larger than the surface reservoir held in the crust and oceans, although this is all a matter of current debate and active research. That said, evidence that there is a nontrivial amount of carbon in the mantle is simply the presence of diamonds. The stable form of carbon several hundred kilometers deep in the mantle is diamond, and every so often these are brought up rapidly and left inside magmatic "intrusions" (magma that gets stuck in the crust). The most well-known such intrusions are called kimberlites, named after the town of Kimberley in South Africa, where they were first discovered. Aside from erupting diamonds, which obviously don't contribute to atmospheric carbon dioxide, the mantle also leaks out carbon dioxide through other volcanism at mid-ocean ridges and, to a lesser extent, hot spots like Hawaii. Therefore, there is a slow and steady replenishment of carbon dioxide to the atmosphere from inside the Earth, and not all of it is drawn out by erosion and weathering; thus, the slow supply of carbon dioxide is enough to help keep the planet's greenhouse blanket in place.

This geological carbon cycle—of extraction of carbon dioxide by weathering and erosion of fresh minerals and resupply by volcanism—is hypothesized, with significant support, to have an important negative feedback, which is the real punch line of this story. (The negative feedback hypothesis, still an active area of debate, is sometimes called the Walker World model, after work done by James C. G. Walker and colleagues, which was similar to the more elaborate BLAG model of my own former Yale colleague Robert Berner and others.) The weathering and erosion of minerals depend on surface temperature in several ways. First, higher temperatures cause more water to evaporate, and thus there is more precipitation when vapor rises up to condense out as rain or snow, which then drives the erosion process. (Precipitation is also helped by the presence of mountains, since winds blow humid air up their slopes to higher altitudes wherein water vapor condenses.) Second, the carbonation or weathering reaction (by which fresh minerals are turned into carbonates) is faster at warmer temperatures. Thus, if excess carbon dioxide is released into the atmosphere by a large volcanic eruption, a huge forest fire, or the extravagant burning of fossil fuels (ahem), the increased temperature by green-house warming causes more precipitation and erosion, and faster weathering of minerals, all of which then draw the carbon dioxide back down. (This drawdown takes millions of years, so it won't save humanity from its profligate ways, unless we can figure out how to make it happen much, much faster.) Likewise, if carbon dioxide levels drop precipitously, as has possibly happened in the distant past (more below), the lack of greenhouse warming causes cooler temperatures, and this limits evaporation, rainfall, erosion, and weathering, which then choke off the carbon dioxide extraction and stop its levels from getting lower; meanwhile, vol-canism slowly releases carbon dioxide, restoring the carbon dioxide levels again. Thus, plate tectonics lets neither carbon dioxide nor temperature get too high or too low, at least for very long (millions and tens of millions of years). In short, the tectonic cycle keeps climate relatively stable over very long times, for hundreds of millions of years. But by stable, we mean it stops variations of many tens of degrees Celsius, but it doesn't prohibit the Earth from falling into a deep ice age or an ice-free sweltering climate.

Life and complex life can evolve and survive moderate swings in climate, but it wouldn't survive catastrophic swings such as a runaway greenhouse effect that would release most available carbon dioxide, evaporate most of the oceans, and turn Earth into something truly hellish, like Venus. Plate tectonics effectively damps out the huge radical swings in climate.

On top of the plate tectonic cycle, the ocean, atmosphere, and ice cover have strong positive feedbacks that drive climate variations. These feedbacks are posi-tive because they amplify (rather than suppress) tiny changes in how we receive energy from the Sun, caused by small fluctuations in the Sun's radiance and in Earth's orbit and spin called the Milankovitch cycles.

The Milankovitch cycles are named for the early twentieth-century Serbian as-trophysicist and geophysicist Milutin Milanković, who hypothesized that changes in the Earth's orbit and spin would induce the observed glacial cycles that last tens of thousands of years. There are three basic effects described by these cycles. The fastest cycle is due to the Earth's spin axis itself slowly twirling around like a top (a process called precession), making a complete loop every 26,000 years; this changes the seasons so that in 13,000 years January will be summer in the Northern Hemisphere. The next cycle describes how the tilt of the Earth's spin axis wobbles between pointing slightly more straight up (that is, more perpendicular to the plane of the solar system) to leaning over a bit more than it does now (currently it is tilted at neither extreme) every 40,000 years; this changes seasonal variations in that more tilt causes colder winters and hotter summers. Finally, the Earth's orbit around the Sun oscillates between being more circular to a bit more elliptical, roughly every 100,000 years; this causes variations in how close the Earth gets to the Sun in its orbit. These cycles combine (along with asymmetry between the Earth's Northern and Southern hemispheres due to different coverage of land and ocean, which absorb sunlight differently) to cause variations over about 20,000, 40,000, and 100,000 years in how much sunlight the planet absorbs, and the signal of these cycles has been verified in the climate record of deep-sea sediments.

The variations in received sunlight caused by the Milankovitch cycles are very small and by themselves would hardly be noticeable. However, positive feedbacks in the ocean and atmosphere amplify these variations and make them large enough to trigger ice age cycles (called glacials and inter-glacials) with periods over tens to hundreds of thousands of years long. Thus, while the plate tectonic cycle calms down big swings in climate, the oceans and ice caps blow things out of proportion, like a bad actor or a science journalist. (Just kidding. Sort of.)

One important positive feedback is due to the oceans' capacity for dissolving an enormous amount of carbon dioxide (much more than is in the atmosphere now but much less than is bound up as carbonates in the crust). However, warm ocean water dissolves carbon dioxide more poorly than cold water, which leads to several important effects.

Imagine if the carbon dioxide concentrations in the ocean and atmosphere were in balance with each other, so that neither is increasing or decreasing at the expense of the other. If the average surface temperatures then rose during one of the Milankovitch cycles, the warming of the ocean would lower its ability to dissolve carbon dioxide, which then would be released to the atmosphere. The additional atmospheric carbon dioxide would enhance greenhouse warming, which would further warm the ocean, which would release more carbon dioxide, and so on. Likewise, if temperatures fall during a glacial cycle, the cooling ocean sucks up more carbon dioxide, which then enhances cooling. In total, the ocean response is a positive feedback that amplifies climate variations. The ocean responds slowly because it takes hundreds to thousands of years for it to fully stir in its surface waters (as mentioned in the previous chapter); however, this is fast enough to react to the much slower changes of the Milankovitch cycles.

While we're on the subject of the ocean's response to warming, it's worth noting the ocean's reaction to carbon dioxide forcing, that is, the release of carbon dioxide from another reservoir (like volcanoes, or burning of biomass and/or fossil fuels). Again, if the ocean's and atmosphere's carbon dioxide concentrations were in balance, but then extra carbon dioxide was dumped into the atmosphere, the ocean would dissolve and draw out some but not all of this surplus down into its depths, mostly through the cold downwelling waters at high latitudes. This process is again very slow because of the ponderous ocean circulation, which is why excess carbon dioxide lingers in the atmosphere effectively for centuries. However, the extra atmospheric carbon dioxide would also eventually warm the ocean, which would then do a worse job of removing carbon dioxide, which in turn would reside and/or accumulate in the atmosphere for even longer. (The biota—namely, plants and trees—also draw down carbon dioxide through photosynthesis, but death and decay release carbon dioxide; biological drawdown of excess carbon dioxide only has a net effect if the total global biomass grows, or if dead biomass gets buried to keep it from rotting, as with fossil fuels, though obviously deforestation and fossil fuel burning negate the effect.)

Another important positive feedback is due to ice caps in the Arctic and Antarctic. Ice cover reflects sunlight back into space, and thus it limits the amount of solar energy the planet absorbs. But if temperatures warm, then the ice melts, less sunlight is reflected, and the Earth's surface warms further, causing more ice to melt, and so on. Likewise, if temperatures cool off, the ice cover grows and reflects more light, causing more cooling and more ice cover, and so on.

Melting of land ice, for example, glaciers and what currently covers Greenland and Antarctica, also causes changes in sea level, which are measurable even today in Earth's rapidly warming climate, with loss of low-lying islands (such as the Maldives in the Indian Ocean). Melting sea ice floating in the ocean has no such effect since it's already in the ocean, although the net change in temperature of the water does cause a modest sea-level change by thermal expansion or contraction. As an extreme example, the melting of both the Greenland and Antarctic ice caps would increase sea level by about 70 meters, or about 200 feet, which would easily inundate most of the world's coastal cities. The loss of ice also possibly has a positive feedback on climate by influencing volcanic degassing. In particular, removing the weight of glacial ice from volcanoes decreases the pressure on their underlying magma, causing it to froth and bubble, like taking the top off a soda bottle, and thus erupt; thus, warming and deglaciation would cause more volcanic carbon dioxide emissions, which would cause more warming, and so on. However, this is a rather new idea (proposed by Harvard geoscientists Peter Huybers and Charles Langmuir) and is still a matter of current research and debate.

The positive feedbacks of the ocean (and its dissolved carbon dioxide) and ice caps amplify any small fluctuation in either heating or cooling. If one of the Milankovitch cycles causes a little extra sunlight to be absorbed by the Earth, then these feedbacks will cause the climate to heat more than would happen just by getting some more sunlight. Likewise, if the Milankovitch cycles cause cooling, feedbacks cause the system to overcool. This overreaction occurs over years and centuries, which is fast enough to amplify a Milankovitch cycle that lasts tens of thousands of years and longer. In this way, we get big enough swings in climate to cause the 20,000- to 100,000-year-long glacial cycles, including the last Ice Age, which ended about 12,000 years ago (and marked the dawn of human civilization).

There have been many variations of climate on Earth, from deep ice ages to globally tropical conditions. I can't cover all of them, but we can review some of the highlights. First, there is evidence that, not quite a billion years ago, before the rise of multicellular life, the Earth was almost entirely glaciated at least once, meaning it was covered in snow and ice in what is called a Snowball event. Stones carried by spreading glaciers from this era have been found in the low-latitude tropics (for example, in ancient geological deposits in Namibia in southwest Africa). This never happened again, possibly because at that time there was the right (or wrong, depending on your perspective) combination of feedbacks to let the climate run away into a globally freezing event.

One hypothesis (which does nicely to illustrate what we've talked about so far) is that the Earth had a major supercontinent, which geologists call Rodinia. Unlike the other famous supercontinent Pangea, Rodinia was centered at the equator in the tropics. When Rodinia broke up, it created more lavas and fresh minerals from rifting (as in East Africa today), and left smaller continents in the tropics with greater exposure to humid coastal environments. The tropics get more solar heating than elsewhere and thus experience most of the world's water evaporation and hence rainfall. This would have caused extremely active erosion and weathering of the post-Rodinian freshly rifted continents, and hence excessive extraction of carbon dioxide. Although this would normally also have caused cooling that would limit rainfall, the tropics don't vary that much in temperature. So even moderate cooling would still have left the tropical continents exposed to a lot of rainfall. As cooling continued, the ice caps grew and amplified the cooling by reflecting more sunlight. Normally, if continents are at higher latitudes, as they are now, ice cover shields them from erosion and weathering and limits carbon dioxide extraction and mitigates cooling. But if the continents are in the tropics, the ice cover is mostly sea ice and so it doesn't shield continents from erosion and weathering. In short, the spread of ice and weathering of rocks proceeded mostly unchecked until the ice caps in the Northern and Southern hemispheres were so large and reflected so much light that there was no stopping their growth, and they almost or completely clapped shut on the equator, encasing the planet in ice for tens of millions of years.

Life survived this calamity in small pockets of water at the bottom of the ocean, and the Earth eventually recovered from this event because (of course) plate tectonics caught up. In short, the global ice cover and freezing temperatures cut off erosion and weathering of rocks and stopped any further extraction of carbon dioxide; however, ongoing volcanism at subduction zones (such as arc volcanoes) and mid-ocean ridges released carbon dioxide and restored greenhouse gas levels. Accumulated volcanic ash probably also helped dirty the ice cover and make it less reflective. Eventually, the surface warmed again and the planet escaped its frozen state. The last Snowball event ended just before the rise of multicellular life, about 600 million years ago, and might have helped trigger the blooming of complex life, which is called the Cambrian Explosion.

Earth has also had major heating events that caused the entire globe to sit at sweltering temperatures, with total loss of ice cover, even resulting in tropical conditions in the Arctic Circle (where fossils of palm trees and prehistoric crocodiles have been found). This occurred most notably about 50 to 60 million years ago, in an epoch called the Eocene, not so long, geologically speaking, after the dinosaurs (who themselves lived in a very warm climate, but not as warm) were wiped out by an asteroid impact in the Yucatán Peninsula. This epoch is associated with high levels of atmospheric carbon dioxide, which were possibly released when massive outpourings of lava during continental breakup and rifting in the North Atlantic through Baffin Bay hit and cooked carbonate-rich ocean sediments. The Eocene is also marked with sharp super-heating events called hyperthermals. One remarkable hyperthermal, called the Paleocene-Eocene Thermal Maximum, hit some of the most extreme temperatures recorded. (Of course, when I say "recorded," no one was around to measure temperature and carbon dioxide levels, but they can be measured indirectly because temperatures affect how the ocean and organisms take up different isotopes of oxygen and carbon, and these isotope levels—really ratios of amounts of isotopes of a given element—are recorded in rocks and fossils; thus, these isotope measurements are "proxies" for temperature, carbon dioxide levels, and so on.)

The Paleocene-Eocene hyperthermal event was very brief and may have been caused by the release of methane from the bottom of the ocean. Even today, microbial life at the bottom of the ocean produces large quantities of methane that are frozen into ices called clathrates. If warming caused by volcanically released carbon dioxide (that is, by cooking sediments) heated the oceans enough to melt the clathrates, then they would release their stored methane. As methane is a very potent greenhouse gas, it would have warmed the climate and hence oceans further, inducing more clathrate melting, and so on, leading to an intense positive greenhouse feedback. However, methane breaks down (due to the high levels of atmospheric oxygen) in less than 10 years and would have been removed from the atmosphere (actually turned into water and carbon dioxide, which are weaker greenhouse gases) relatively rapidly. These characteristics of methane release and loss possibly explain why this thermal maximum was so intense but also so brief.

Since the Eocene, the Earth has undergone a continuous cooling event for about the last 50 million years. During the Eocene, Australia and Antarctica were connected into one continent whose coastline then protruded north toward the equator into warm latitudes. Ocean currents hugging this coastline would carry warm water from these temperate climates to the polar regions of Antarctica, and thus keep it warm and ice free. But then Australia broke away from Antarctica and headed north toward Asia. The Antarctic coastal currents then stayed confined to cooler polar waters, as they do today, in what is called the Circumpolar Current, and stopped delivering warmer water to Antarctica. This caused enough cooling of Antarctica to start ice growth, which then triggered more cooling that promoted more ice growth, and so on.

The northward migration of the Australian subcontinent was part of the same plate tectonic movement that led to India colliding with Asia and pushing up the Himalayas. The appearance of this giant mountain range also possibly led to a site for extensive erosion and weathering and carbon dioxide removal, which is referred to as the Raymo-Ruddiman hypothesis (after American paleoclimatologists Maureen Raymo and William Ruddiman). Mountain ranges tend to trigger precipitation as moist air is pushed upslope to higher and cooler altitudes. Moreover, the warming of continents during the summer causes convective upwellings that draw in moist air from the oceans, which then leads to enhanced rainfall and snowfall over the continents (this is called the monsoonal circulation). Overall, the enhanced precipitation over the newly built Himalayas led to more erosion and weathering, and hence greater extraction of carbon dioxide, which thus amplified the long cooling trend.

However, during the 50-million-year cooling period, Antarctica lost its ice again about 30 million years ago and regained it about 15 million years later in an epoch called the Miocene. Since the Miocene, the Earth has had ice caps, and the details of our climate history in the last few million years are easier to see because there are more temperature "proxies" to measure, in ice cores, tree rings, cave deposits, and the like. In the last several millions of years, there have been repeated short ice age or glaciation events, lasting tens to hundreds of thousands years or so, which coincide well with the Milankovitch cycles, as discussed above.

The last major ice age event is called the Pleistocene, which lasted in total (with some shorter ups and downs) from about 2.6 million years ago to about 12,000 years ago, at the dawn of civilization. But even since the appearance of humans, about 7 million years ago (we'll get to that later), the Earth has been in a deep cooling trend and essentially in an ice age, and thus has had ice caps for all of humanity's existence. In short, we are ice age creatures who did not evolve to live in Eocene-like conditions. To a large extent, this is why the loss of our ice caps in Antarctica and Greenland would be so catastrophic: apart from the resulting drastic changes in sea level, we would find ourselves on an Earth our species has never known and was never meant to inhabit.

The causes of a stable climate and habitability as well as mechanisms for climate variation are important lessons for humanity, which faces its own complex crisis regarding human-induced climate change. Greenhouse warming due to anthropogenic carbon dioxide emissions from burning of fossil fuels is not a subtle effect and was reasonably well predicted more than 100 years ago by Swedish physicist and chemist (and Nobel laureate) Svante Arrhenius. And that atmospheric carbon dioxide levels have been (and still are) increasing at an unnatural rate was verified by American geochemist Charles Keeling from nearly 60 years of measurements from the top of the Hawaiian volcano Mauna Loa. The precise details of how the climate will warm in response to carbon dioxide emissions are still debated, but that it will warm significantly is not.

But there was (and to some extent still is) a debate about whether our current release of carbon dioxide is triggering climate change, or if the climate would change on its own anyway, in which case human activity is inconsequential. The Earth's climate has varied on its own, but usually when carbon dioxide levels have changed. Indeed, our climate has even responded drastically when large and normally stable reservoirs of carbon were burned or liberated suddenly by, say, volcanic events. So there is no sensible reason to think Earth's climate will respond differently if we also release vast quantities of carbon dioxide from another huge carbon reservoir. Wondering if our activities make a difference among all the natural causes of climate variations is like asking whether playing Russian roulette during a gun battle will affect your chances of survival. If the goal is not to die, then the answer is don't play Russian roulette.

But in the end, addressing anthropogenic climate change is not really about saving the planet; it's about saving us—that is, preserving a narrow habitable zone that is suitable for us (and a few other) ice age creatures. Regardless of how bad we make things for ourselves, the planet itself will be just fine in a few million years; plate tectonics will eventually draw down all our waste carbon dioxide and things will go back to business as usual. That we can't wait that long is not the Earth's problem.

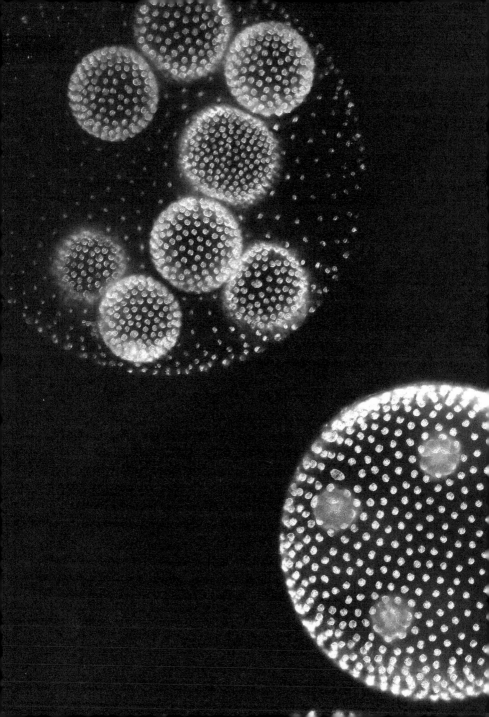

7 LIFE

The origin of life—that is, how life began—is one of the Holy Grails, and still one of the great unanswered questions, of natural science. It is more appropriately, if less poetically, called the abiotic origin of life, the formation of life from inorganic, inanimate matter.

Before we can try to pinpoint the dawn of life, we have to have a definition of life, so we know what to look for. (Although the signs of life might be intuitively obvious, it's not really scientific to say—abusing the words of Supreme Court justice Potter Stewart—"I know it when I see it.") In its simplest sense, life is a chemical reaction that directly or indirectly takes matter and energy from its surroundings to expand, propagate, and reproduce itself. Such a reaction is called autocatalytic in that the products of the reaction facilitate or even accelerate the reaction itself. In the photosynthetic reaction, plants use the energy from sunlight to combine water and carbon dioxide to make long sugar molecules that comprise much of the plants' bulk (in the form of cellulose), which then supports more photosynthesis. Alternatively, aerobic, oxygen-breathing cells and animals like us eat these plants (or eat other organisms that eat these plants) and use their matter and stored solar energy to make more cells that eat more plants. By reproducing and propagating itself, life then actively spreads or seeks out sources of matter and energy.

Some of the characteristics of life could apply to a number of nonliving chemical reactions—for example, fire. Like aerobic life, fire consumes matter and energy, and reverses photosynthesis to make water and carbon dioxide. Like life, fire spreads outward to consume fuel (for example, wood and grass) and catalyzes itself by heating up fuel until it ignites.

However, two other definitions of life set it apart from fire. Life's reaction doesn't just consume matter, it also makes complex molecules that form a template with which to catalyze more molecules like themselves and hence reproduce. Not only does such reproduction cause faster reactions, it allows for the inheritance of information from previous molecules. In contrast, fire doesn't duplicate complex molecules, it just makes simple ones (water and carbon dioxide). Second, life evolves by natural selection: if the environment becomes unsuitable for sustaining the chemical reaction, life can possibly adjust (if the change in environment isn't too fast). But this adjustment relies on the duplication of previous organisms being imperfect, such that, as new generations of organisms perpetuate, they develop some variety and are not exactly clones of their progenitor. Thus, the group of organisms, or species, will have enough diversity so that some members will be better suited for—and survive—an adverse change in the environment, while the ill-suited members die off; this is the essence of Darwinian natural selection. Fire cannot make such adjustments—if the environment gets too cold or wet, then it just stops; there is no selection whereby some fires that are better suited to cold or damp perpetuate while others die off. In short, one can say that life is a self-sustaining, energy-consuming chemical reaction whose product molecules catalyze or reproduce themselves, but there is enough variety in the products that evolution by natural selection can occur if the environment goes sour (slowly enough). Okay, that wasn't so short.

All life on this planet is cellular in that the chemical reaction sustaining life happens inside a capsule. This capsule is enclosed in a slightly leaky membrane that lets in nutrients and sources of energy and also protects the reaction from being dissipated or wiped out by, say, a crashing ocean wave. It is even possible that the earliest such capsules used bubbles inside of lava rocks, like pumice, for protection. Some noncellular entities, like viruses, which are free-floating genetic material wrapped in protective sheaths, have some characteristics of life such as natural selection but cannot reproduce themselves without hijacking the cellular machinery of other organisms; thus, whether they are life or not is still subject to debate.

The earliest verifiable fossil of life is of single-cell microbes (like bacteria) from about 3.5 billion years ago. There may have been earlier life, but the fossil finds are controversial. As much variety as there is to life now, the basic chemical components for making life have hardly changed in those nearly 4 billion years, and in fact one needs only a handful of basic elements to make the key components of life.

The first essential elements for life are hydrogen, carbon, and oxygen, which life invariably gets from all the ubiquitous water and atmospheric carbon dioxide; not only are these necessary to make various sugar molecules that provide plant structure or an aerobic energy source, but the sugar molecules also contribute part of the backbone for genetic material, RNA (ribonucleic acid) and DNA (deoxyribonucleic acid), the blueprints that allow complex molecules of life to reproduce themselves. Moreover, when these sugars are "reduced" by having the oxygen removed (in general, reduction means a gain of electrons, usually the ones that oxygen has been holding onto in an oxide compound), they leave just hydrocarbons in the form of fatty acids that make up, for example, the lipids in cell membranes as well as fat cells that store these oils as a compact form of energy. Carbon and oxygen are also used prodigiously in other important molecules, which we'll get to shortly.

The next important element is nitrogen, mostly in the form of the amide ion, which has one nitrogen, two hydrogens, and a spare electron, with a negative charge, that is used to latch onto other atoms (or groups of atoms) to make amine molecules. Amides are ultimately derived from ammonia—a molecule composed of a nitrogen and three hydrogen atoms—by lopping off one hydrogen. The amides latch onto another carbon-oxygen-hydrogen molecule (a carboxyl compound rather than a sugar) to make amine molecules called amino acids, which are the basic building blocks for proteins. Proteins are important because they are very diverse in their use and make sundry ingredients from enzymes to muscles. Enzymes are particularly important for accelerating (that is, catalyzing) chemical reactions—like breaking down molecules in food—so that they are fast enough to sustain biological activity. Also, with a little electrical or chemical stimulus, proteins fold and twist into various forms, making them useful for mobility—like the waving flagella that help bacteria swim or our own muscles—and a little mobility goes a long way for seeking food and energy sources.

Nitrogen also combines with carbon, oxygen, and hydrogen to make compounds called nucleobases, which are crucial components of the nucleic acids DNA and RNA. These nucleobases are adenine, cytosine, guanine (all of which occur in both DNA and RNA), thymine (only in DNA), and uracil (only in RNA), and show up in diagrams of DNA and RNA as the A, C, G, T and/or U "bases" that make up the rungs on the helical ladders of DNA and RNA (RNA looks like half a ladder, cut lengthwise).

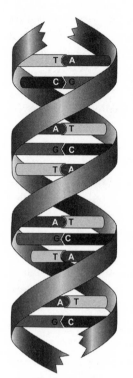

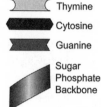

Adenine

Thymine

Cytosine

Guanine

Sugar
Phosphate
Backbone

The deoxyribonucleic acid (DNA) molecule is made of stacked nucleotides, each of which is a combination of a sugar-phosphate group attached to one of several nucleobases, adenine, cytosine, guanine, and thymine. The DNA is then shaped like a helical ladder, wherein the nucleobases form the ladder rungs, and the sugar-phosphates connect together to make up the ladder sides. The nucleobases form sequences holding genetic information and cellular instructions, and also link across the ladder in specific combinations (as indicated), which allows the DNA to reproduce itself faithfully after splitting. (Courtesy Barbara Schoeberl, Animated Earth LLC.)

Finally there is phosphorus, which appears only bound with oxygen as a phosphate (a phosphorus atom bound to four oxygen atoms). Phosphates bind with a sugar and the different nucleobases to make compounds call nucleotides, which stack up to make the helical half or full ladders of RNA and DNA. In particular, the sugar plus phosphate parts of each nucleotide stack like vertebrae (actually, the sugar end of one nucleotide bonds with the phosphate end of the next one) and make the ribose (for RNA) or deoxyribose (for DNA) spine or ladder sides, while the nucleobases stick out like ladder rungs. (Hence you see why RNA stands for ribonucleic acid and DNA for deoxyribonucleic acid.) The nucleotides also make molecules that store and carry energy such as adenosine triphosphate (ATP), which is the gold energy currency in cellular machinery since it carries three phosphates that are highly reactive. Phosphates and nitrogen also combine with fatty acids to make the phospholipids in cell membranes.

In DNA and RNA, the nucleobases, or just "bases," chemically bond with one another, but only in specific complementary ways. For example, the nucleobase A binds only with T, and C only with G to complete either side of the DNA ladder; thus, one full DNA rung will be A on one side of the ladder and T on the other, and so on. During cellular reproduction, the DNA splits lengthwise, and the bases sticking out as broken rungs on each half ladder attract their complementary bases floating loosely in the cellular soup, and thus reconstruct the other side of each ladder, and hence the DNA reproduces itself. It is this feature that makes DNA a template for its own reproduction, a self-duplicating molecule, and hence the core feature of all life, or at least life as we know it. DNA also holds genetic information for both reproduction and running cellular machinery, and this information is coded or written in the sequence of base-pair ladder rungs. In addition to reproducing itself, the DNA can split and copy segments of its split strands to RNA (again by nucleobase matching), which is then sent off on errands, like organizing amino acids into specific proteins for different tasks.

In the end life is fundamentally made up of four basic compounds (other than water)—sugars, fatty acids, amino acids, and nucleotides—and these are built on just five elements: hydrogen, carbon, oxygen, nitrogen, and phosphorus. While most hydrogen was made in the Big Bang, the other four were made inside of stars. Other elements in lesser quantities come into play, depending on the various organisms: for example, iron in our blood carries oxygen, which is used for converting sugars for our energy needs. But these four basic compounds, made from only five elements, are what all life has in common, and to make life as we know it on Earth from scratch, one needs to have these building blocks.

One of the most famous attempts to create the building blocks of life from inanimate matter was a set of experiments done in the 1950s by University of Chicago chemistry graduate student Stanley Miller and his eminent adviser Harold Urey (who was already renowned for many other accomplishments, including the discovery of deuterium, an isotope of hydrogen, for which he received the Nobel Prize in chemistry in 1934). Miller created a mixture thought to represent Earth's primordial atmosphere, containing hydrogen and compounds made with hydrogen, like water, methane, and ammonia. He then exposed the concoction to high temperatures (basically steam) and electrical shocks; after a few days, the flask contained several amino acids. However, the experimental atmosphere was really more typical of the pre-solar nebula and what one might find in the outer solar system in Jupiter, Saturn, and some of their moons. Such an atmosphere was likely cooked and blown out of the inner solar system during its formation, and the Earth's first atmosphere was probably dominated by carbon dioxide and water from volcanic degassing, thus nothing like that in the Miller-Urey setup. Nonetheless, the pioneering Miller-Urey experiments showed that simple reactions among a few compounds could result in at least one of the essential building blocks of life. This work also paved the way for decades of related experiments seeking life's building blocks by simulating prebiotic soup in primordial atmospheric and oceanic conditions. Indeed, amino acids could even form in environments in space; several amino acids (not all the same as on Earth) were found on the Murchison meteorite, which is a carbonaceous chondrite from the Asteroid Belt. Whether meteorites seeded Earth with amino acids is unknown and perhaps irrelevant since they form in a wide variety of environments anyway, and the other building blocks of life also need to be produced.

Not long after the Miller-Urey experiments, Spanish biochemist Joan Oró was able to form nucleobases—which, recall, are the rungs on the DNA and RNA ladders—as well as amino acids. Forming full nucleotides (which stack to make the full RNA and DNA molecules) has proven more elusive until recently. In the last decade there has been considerable progress in synthesizing various building blocks including lipids, amino acids, and nucleotides from compounds thought to be on early Earth, especially in work led by Cambridge chemist John D. Sutherland. The simplest possible cells are made of a DNA strand encased, along with a nutritious soup, inside a fatty acid or lipid bubble or membrane to make a cell wall; and indeed various recent experiments (led by Harvard biochemist Jack W. Szostak) have found that the right lipids can spontaneously form bubbles that allow in nucleic acids and make something like a proto-cell. Thus, researchers have made great progress in getting close to "spontaneous" abiotic cellular formation since Miller and Urey's experiments 60 years ago.

When and where did life first form on our planet? Although the oldest microfossil is about 3.5 billion years old, life probably had some progenitors leading to that organism, perhaps through millions of years of false starts and variations. It is even possible that earliest life was based on the simpler RNA molecule, rather than DNA, as the backbone of biological reproduction. In modern cells, RNA acts only as an errand runner for DNA, for example, to make specific proteins. However, biochemists Sidney Altman (from Yale) and Thomas Cech (University of Colorado) demonstrated that RNA could catalyze or reproduce itself, a discovery that earned them the Nobel Prize. This finding gave critical support for what is called the RNA World hypothesis, that is, that earliest life was based on RNA's simpler method of reproduction, which was the precursor to the more complex reproduction carried out by DNA in all cellular life today.

Charles Darwin, like Miller and Urey, thought life arose at the Earth's surface, in pools of primordial soup that had the basic ingredients for self-generating life, which then drew on the Sun's energy through photosynthesis (and indeed the first verifiable life was a photosynthetic microbe). But if life was forming this way prior to 3.5 billion years ago, it would have had a very rough go of it. During this time the surface was an extremely hostile environment, likely still very hot from a carbon dioxide atmosphere, still highly volcanic, given the Earth's very hot interior, and most likely pummeled by stray asteroids, especially during the Late Heavy Bombardment period from 4.2 to 3.8 billion years ago. Thus, the surface was not likely to be hospitable to the formation of the first fragile life forms.

In the late 1970s, hydrothermal vents and chimneys at the Galápagos mid-ocean ridge (where two of the Earth's bigger tectonic plates spread apart) were discovered by Oregon State University geologist Jack Corliss and colleagues (using the deep-sea submersible *Alvin*) to be teeming with life, but at the bottom of the ocean, far from any sunlight. At these vents, extremely hot water (above the boiling temperature at sea level, although not boiling at those deep submarine pressures) circulates through these volcanic ridges and rises up as hot plumes of water loaded with minerals and dissolved volcanic gases like carbon dioxide, hydrogen, and hydrogen sulfide. Inside of these super-hot plumes of water were discovered microbial organisms called archaea, which are vaguely similar to bacteria; these specific ones were thermophiles, meaning they love hot water. Entire ecosystems of larger organisms like tube worms were found around these vents, feeding off the nutrients sequestered by the archaea and bacteria thriving in this environment. For example, tube worms obtain energy and nutrients from bacteria, which are in turn powered by chemosynthesis (rather than photosynthesis, given the dearth of sunlight), wherein hydrogen sulfide from the vents is used to make organic carbon by stripping it off of carbon dioxide molecules. This discovery showed life flourishing far from our favorite energy source, the Sun, and surviving off the heat and chemicals coming from the mantle and crust. This suggested that earliest life might have formed at the bottom of the ocean, protected from the hostile surface environment, living off the weak but reliable mantle energy source. This also implied that life could form on planets (like Jupiter's moon Europa, perhaps) that are not at the right distance from their star to have liquid water; it could be supported by volcanic energy, provided it was enough to sustain liquid water.

The archaea first discovered at these vents, and later in many other environments, including unexpected ones like hot springs, acid pools, salt flats, polar ice, and even our own intestines, were first thought to be bacteria because, like bacteria, they consisted of some simple DNA strands wrapped in a lipid bubble. However, archaea and bacteria were then found to have more differences than similarities, for example, in their RNA, energy use (their metabolism), the chemistry of their cell walls, and the flagella they use for swimming. However, bacteria and archaea are both prokaryotes, meaning they have a simple cell structure and rarely ever occur as cellular colonies, and never as multicellular life.

The appearance of life on Earth's surface relied and still relies on photosynthesis. The emergence of photosynthesis was one of the greatest biological revolutions on the planet, perhaps second only to the first appearance of life, given that it is the foundation for powering almost all biology on Earth (directly or indirectly) with abundant solar energy, in addition to having radically altered the atmosphere. How photosynthesis works is still an active area of research, and while I have repeatedly simplified it (up to now), the reaction is rather complicated and involves a few steps. Typically a photon from sunlight is captured in a cell by proteins containing pigments like chlorophyll, and the photon's energy is used to split a water molecule and strip off an electron, leaving a hydrogen nucleus (a proton) and an oxygen, which is released as a waste product. The freed electron is basically an energy token and it's used to create cellular energy carriers like ATP. Some of this stored energy is then used to take atmospheric carbon dioxide and effectively swap one of its oxygens with two hydrogens to make the final product, sugar (and more waste oxygen). Production of sugar effectively reduces the carbon to organic form by letting it hang on to more electrons rather than share them with the electron-greedy oxygen (and since hydrogens are smaller and less greedy). The more oxygens removed this way, the more reduced the carbon becomes (more on this later) and the more energy it stores with its own stash of electrons.

Thus, one of the first dominant forms of microbes that made it to the surface was photosynthetic bacteria, which were much like cyanobacteria (often called blue-green algae, although this is a misnomer since real algae are not bacteria). These bacteria made microbial mattes, sheets of microbes that would pile on top of each other to keep exposed to the Sun; they would then harden and calcify (or make carbonates) to form stromatolites, the oldest verifiable fossils. Being photosynthetic, these microbes converted carbon dioxide and water to sugar and released free oxygen as a waste product. Oxygen is a very reactive gas in that it (as alluded to already) wants to steal electrons from other elements by binding with them, and so tends to react and bind with almost any available element except ones even more reactive (or electron hungry) than itself, like chlorine or fluorine. For many life forms, oxygen is caustic and poisonous, and exposure to it leads to chemically reactive "burning," much as if we were exposed to chlorine, one of the first poisonous gases used during World War I.

At first, the photosynthetic waste oxygen did not accumulate in the atmosphere and was extracted by reacting with elements such as iron and iron-rich minerals at the surface and in the ocean to make iron oxide, that is, basic rust. Eventually the available iron was used up, leaving behind lot of ancient geological deposits of iron oxide (called the Banded Iron Formations, which comprise many of the sources for iron mining today) after about 2 billion years. After that, without minerals and metals to buffer it, oxygen accumulated to about the levels it has today, about 20 percent of the atmosphere.

The leveling off of oxygen can be simply attributed to its reaching a balance with all the organic material produced (sugars, fats, methane gas, and so on), which reacts with oxygen again eventually to give back carbon dioxide and water. In chemistry it means the reaction reached a steady state, or the production of oxygen by photosynthesis is matched by its consumption in the reverse reaction. As noted already, one way to reverse the photosynthetic reaction and reach that balance is with fire, which releases the stored solar energy as heat and light. Another way is with aerobic organisms (like us), who consume sugar and fat, react it with oxygen, use the released solar energy, and expel carbon dioxide and water. The progenitors of aerobic organisms were similar to bacteria and evolved to use the oxygen to consume their own sugar energy source as a contingency for when solar energy was lacking. This skill would come in handy later (more below). In the end, this eventual balance between photosynthesis and aerobic consumption is marked by mostly flat oxygen levels.

However, the amount of atmospheric oxygen is enormous—again, being 20 percent of the mass of our atmosphere, or about a quadrillion metric tons (10^{18} kilograms). There is therefore also a huge reservoir of organic matter that complements all this free oxygen, that is, the other product of the photosynthetic reaction—namely, again, sugar, although it is usually just referred to generically as organic carbon. (In contrast, carbon that has been bound up in rocks as carbonates by weathering reactions is inorganic carbon.) Much of this organic matter is hidden away from the atmosphere or else it would eventually react with free oxygen by various means. On Earth, such sequestering of organic carbon is easily done by burying it at the bottom of our huge oceans or under sediments that are constantly produced by the erosion of perpetually building volcanoes and mountains. This accumulated organic matter reservoir is today several thousands of times bigger (in terms of mass of carbon) than that of the extant biosphere, which is itself, then, a relatively tiny system in which oxygen is now continuously produced and consumed at about the same rate.

How aerobic organisms use sugar to gain energy during respiration deserves a few words. When sugar (or a hydrocarbon) is simply burned by reacting it with oxygen, the organic carbon's stash of energetic electrons (built up in the photosynthetic reaction discussed earlier) is repossessed by oxygen and dropped into a low energy level in the oxygen atomic structure or "electron valence shells," thus releasing energy as heat and light. Alternatively, when the sugar is used by an aerobic organism, metabolic reactions cause the organic carbon's electrons to be slowly leaked back to the electron-greedy oxygen and used to build up electrical voltage, whose energy is (eventually) stored in the creation of ATP, which then powers cellular machinery. During this whole process, a fraction of the sugar's stored energy does not get used to make ATP but is instead released as heat, which is what keeps warm-blooded creatures warm. In either case, of burning or aerobic sugar consumption, once the oxygen gets its electron fix, it leaves with the waste carbon dioxide and water.

As noted in chapter 5, nitrogen makes up most of the other 80 percent of the atmosphere, which basically provides a reservoir for many of the biological building blocks discussed above. However, nitrogen gas is relatively inert and not easy to react with and thus capture or "fixate." It takes a lot of work, mostly from bacteria and archaea in the oceans and soils, to make, for example, ammonia that then gets used to make amino acids by larger organisms like plants; we ourselves do not get nitrogen from the air. (However, we owe the prolific production of fertilizer for present-day agriculture, which sustains the world's massive population, to synthetic nitrogen fixation from the atmosphere, a process that was discovered more than a century ago by German scientist Fritz Haber, for which he was awarded the Nobel Prize in chemistry.)

For most of the Earth's first billion years, the biosphere was dominated by simple single-cell prokaryotes, that is, bacteria and archaea. Complex cells—such as what make up animals and plants and various complex single-cell organisms like fungi, amoebae, and paramecia—arose about 2 billion years ago, and these are called eukaryotes. These cells are very different from prokaryrotic cells in that a typical eukaryote has a membrane held up by a cytoskeleton, a nucleus holding the DNA instead of letting it float freely, and various pieces of machinery to run the cell called organelles. These cells can also change their shape and have the membrane engulf and eat other organisms. How did these organisms arise?

The most widely accepted hypothesis for the origin of eukaryotes is called endosymbiosis, which involves two prokaryotes initially joining, perhaps by one eating the other or one invading the other (and it doesn't matter which since it would be hard to tell the difference). The process could have involved archaea "engulfing" bacteria or vice versa. However many times this might have happened, the combination became a symbiotic exchange. For example, aerobic bacteria, which are able to remove oxygen and use it with sugar for energy, would have been useful partners to archaea for which oxygen was poisonous. Alternatively, photosynthetic bacteria inside larger cells would generate sugar for their host. Diverse symbiotic combinations like this could have given a large evolutionary advantage in a changing oxygenating environment, and so the eukaryotes got a foothold.

Organelles, some of the machinery inside eukaryotes, are thought to have formed from this symbiotic partnership. The proof of this is that our very own human cells contain organelles called mitochondria that look pretty much exactly like bacteria, with their own small DNA strands, and they are largely responsible for converting energy inside our cells. Plants likewise have bacteria-looking organelles called chloroplasts that perform photosynthesis. Either way, the symbiotic combination was well suited for exploiting the rising oxygen levels on the planet along with all the sugar and lipids accumulated by photosynthetic bacteria. Indeed, sugars and fats are much more efficient and transportable energy sources than one obtains from sitting around absorbing sunlight all day. So marks the origin of conspicuous consumption. The fact that now we can not only store sugars and fats to power our bodies but also store them in automobiles and airplanes to increase our mobility, is more or less the same principle.

Since eukaryotes are in effect combinations of other cells, they are naturally bigger than prokaryotes, and can even get much, much bigger. Eukaryotic size is not necessarily limited because its machinery is all throughout its interior, so if it gets bigger it also has proportionally more organelles. Prokaryotes are thought to have hardly changed size (or shape) in nearly 4 billion years, primarily because most of their cellular machinery is on their enveloping cell membrane as tubes and pumps for chemicals, while the interior is just a chemical soup and some free-floating DNA. If they got bigger, then all the expanded membrane and its machinery would have to support an even bigger interior volume; if the cell doubled in radius, the surface machinery might quadruple but the interior volume would go up by a factor of eight. Eventually the cell's surface would be unable to keep up with its volume, and so increasing prokaryote size is a disadvantage.

Eukaryotes also diversified more than prokaryotes because of differences in reproduction. Prokaryotes primarily undergo cell division or mitosis and just clone themselves and so have hardly changed in nearly 4 billion years. Simple single-cell eukaryotes undergo cell division but also will subdivide and shuffle their own nuclear DNA and share some of it with a partner through meiosis and sexual reproduction. The advantage of shuffling and exchanging is that they both drive diversity and dilute possibly fatal genetic mistakes caused by damaged segments of DNA (which are more likely to be lost in a shuffle, but will certainly not get lost if the

DNA is just cloned whole cloth). Both diversity and error control would give a further evolutionary advantage and perpetuate these characteristics.

Multicellular animal and plant life likely first occurred by formation of cellular colonies. A colony is characterized by all the cells being identical, while a multicellular organism has specialized cells that serve different roles (such as our own muscle, brain, bone, and eye cells). While prokaryotes can form crude colonies of filaments and microbial mattes, single-cell eukaryotes can form various and diverse colony structures like the volvox (floating spherical colonies of algae, shown in the chapter opening) and slime mold. The step from a colony to a multicellular organism was probably reasonably straightforward given the diverse ways eukaryotes could adjust and evolve. Cells on the surface of a colony would be more responsible for absorbing energy and nutrients from the environment, while deeper cells would pump nutrients and water to the interior (thus setting up an early circulatory system). The different environments inside a colony would in essence drive divergent evolution of its cellular members to become different and specialize. Eventually cells specializing in, for example, moving the whole conglomerate organism around or sensing predators and prey might provide evolutionary advantages for specific environments.

The rise of multicellular life, however, took a very long time on Earth. Prior to about 640 million years ago, the biosphere was still dominated by single cells. From about 640 to 540 million years ago, there were some crude frond- and tube-shaped life forms (an era called the Ediacaran), but these went extinct. Then, about 540 million years ago, multicellular life bloomed in diversity with a huge number of different marine creatures, most of which you would probably not recognize except to think they looked like very ugly scorpions, centipedes, and crabs.

This event was called the Cambrian Explosion, and in essence it marks the beginning of the fossil record because so many of these creatures had hard parts, shells and skeletons, that could be preserved without decaying away. Of course this means we might have missed many fossils from earlier that did not have hard parts, but even today modern paleontology is able to discern the presence of life by smears of biological and genetic material left on rocks by long-gone invertebrates. Moreover, sediments from before the Cambrian Explosion—now cemented together as rocks—show few signs of being disturbed by digging, squirming creatures (an effect called bioturbation), but after the event such signs are pervasive in the sea-sediment record.

The rise in shelled creatures, whose hard parts are made from carbonate minerals, might have come about due to the volcanic accumulation of atmospheric carbon dioxide that got Earth out of its Snowball state (as discussed in chapter 6). In particular, the heightened carbon dioxide levels would eventually have dissolved in the ocean and been removed by mineral weathering, providing material for shells. Thus, the end of the Snowball event might have triggered the Cambrian Explosion. In the last 400 million years or so, plants and animals colonized the continents and continued to evolve and diversify to fill every niche and nook and cranny they could. However, note that the time since the Cambrian Explosion—called the Phanerozoic—is only about 10 percent of Earth's entire history; most of Earth's biological history was solely occupied by microbes.

During Earth's long biological history, a lot of solar energy was effectively stored as sugar and fats and other organic material at the same time as the massive amount of atmospheric oxygen was accumulated. As noted already, most of this organic matter was buried and hidden from the oxygenating surface environment under sediments and at the bottom of the ocean. A tiny fraction of this hidden organic matter that was buried deep in the crust was subjected to just the right high temperatures and pressures to make various kinds of fossil fuels, in effect by slowly cooking the sugar molecules to remove oxygen and create even more reduced carbon (again, reduction means gaining electrons that the oxygen had been holding on to). Buried marine organic material undergoing this process would make oil and gas hydrocarbons (molecules of hydrogen and carbon but no oxygen); some of these reservoirs would then get pushed up by plate tectonics, or exposed when sea level fell, to be in continents again, such as the western United States from Texas to Wyoming, which was a sea during the time of the dinosaurs. Buried terrestrial organic material like trees and swamps would, if cooked at the right conditions, eventually turn into coal, which is more or less pure carbon (swamps also produce peat, which is an intermediary step to coal but is also considered a fossil fuel). Oil, gas, and coal (and peat) combined make up our reservoir of fossil fuels, although the vast majority of it, about 85 percent in terms of carbon mass, is coal.

Most coal was produced about 300 million years ago during a geological period called (for good reason) the Carboniferous, which was not so long, geologically speaking, after plants colonized the continents.

There are about 4 trillion metric tons of carbon buried as fossil fuels in total, which is twice as much as the carbon mass extant in the entire biosphere today (both living and dead biomass). However, of all the organic matter created as a complement to atmospheric oxygen, most of it, nearly 15 quadrillion metric tons of carbon—about four thousand times as much as in fossil fuels—sits in the crust untransformed to fossil fuel and is far too hard to extract or use. This is generically termed kerogen. Kerogen is itself one of Earth's major reservoirs of carbon but is, to give some perspective, still about one-fourth the size of the inorganic carbon reservoir stored as carbonate minerals in the seafloor and continents. Carbonate minerals and kerogen together have sequestered almost all of Earth's primordial carbon dioxide that was once in the atmosphere, and thus they keep our climate from becoming something like that on Venus. Because the kerogen reservoir is so huge, when only a tiny fraction of it hits just the right temperature and pressure conditions, it still yields a giant fossil fuel source.

In the end, the forms of carbon and hydrocarbon in fossil fuels are even better fuel sources than sugar since all their oxygen is removed—hence, there is even more material with which oxygen can react. In a sense, fossil fuels represent not just stored solar energy captured by photosynthesis but also geothermal energy used to reduce the sugars to their unoxygenated state. Even forgetting about all the biological production before the Cambrian, there are hundreds of millions of years and trillions of tons of this stored energy available to consume. The availability of this cheap, concentrated, and transportable source of stored energy has transformed human civilization and given rise to innumerable technological and social advancements. But then, the extreme utility of this resource has compelled us to burn it at astounding rates, going through millions and millions of years of stored carbon in a matter of decades, leading to the current impact it has on the habitable climate and environment—that is, habitable for us.

8 HUMANS AND CIVILIZATION

The origin of humans occurred hundreds of millions of years after the rise of multicellular life. Thus we are hardly doing justice to all the life forms that came in between, especially everyone's favorite: dinosaurs. So, by way of a little backstory (though not much), our mammal ancestors existed even during the time of the dinosaurs, although these were small rodentlike creatures that filled ecological niches not already taken by the dominant dinosaur species (for example, by being nocturnal and living underground). After the demise of the dinosaurs, around 65 million years ago, triggered by an asteroid impact in the Yucatán Peninsula, mammals found more niches and grew in size and diversity. The largest mammal, known variously as the Indricotherium, Paraceratherium, or Beast of Baluchistan, lived over 20 million years ago in Central Asia. This beast was a hornless predecessor to the rhinoceros but far more massive and, with its long neck, resembled the even more massive sauropod dinosaurs like the Brachiosaurus, as do (for that matter) modern giraffes.

Some smaller mammals that were contemporaries of the dinosaurs had long found a special niche in trees, which were a haven from larger animals and provided unique food sources such as leaves, fruits, and insects in the tree crowns and the canopies of rain forests. A rodent-like tree shrew was probably the common ancestor to all primates, which first arose around the time the dinosaurs went extinct, if not earlier. These first primates were distinct in having prehensile thumbs and toes, unique skeletal structure around the eye sockets, and a predilection for fruit. Primates developed and existed in a common form in Africa, East Asia, and the Americas until tailless apes branched off from the rest of the Old World tailed monkeys (like the baboon, lemur, rhesus macaque, and so on) around 30 million years ago.

Within the family of tailless apes, the great apes and lesser apes (the gibbon family) separated from each other about 18 million years ago. The family of great apes first spawned the genera (the branch of life above species) of the orangutan, then the gorilla, and finally the chimpanzee and human, who separated from each other probably around 7 million years ago. Those four genera comprise now the existing great apes, each of which, other than humans, has two species (for example, the chimpanzee genus is really called *Pan* and includes the common chimpanzee and the bonobo species); humans have only one species left, *Homo sapiens*—that is, us.

In general, the great apes spend much less time in trees than their ancestors did and are (as the name implies) bigger and generally more intelligent, at least from our perspective. Leaving the trees freed up hands and thumbs, and the resulting ability to manipulate the environment (for example, using a stick to get food) gave some evolutionary advantage in the environment in which they lived. Still, none of these great apes climbed to the top of the food chain until humans started to get an upper hand (so to speak).

What drove this development and speciation of primates over a few tens of millions of years? Recall that, for the last 50 million years, the climate had been steadily cooling. But despite the overall cooling trend, the planet warmed from about 30 to 15 million years ago, and the resulting tropical conditions led to rain forests in Africa and Asia and hence a suitable environment for tree-dwellers. However, after 15 million years ago, the cooling and drying trend resumed, which likely diminished arboreal habitats.

Plate tectonic motions also added to the environmental change. Continental drift was already responsible for altering ocean currents and uplifting the Himalayas (see chapter 6) and thus driving the entire 50-million-year cooling trend. By 30 million years ago, the African continent was colliding with Eurasia and slowly closing the seas, called the Tethys and Paratethys, between them; this caused gradual loss of coastal tropical ecosystems and replaced them with drier continental ones. Moreover, the East African Rift valley (and the Red Sea Rift) started ripping open during this time, leading to elevated terrain because the hotter crust and mantle were buoyant and bobbed up. (East Africa is still rifting or tearing apart, which will lead to continental breakup and the formation of a new seaway in tens of millions of years.) Elevated land would lead to drier and cooler conditions, and much of the

African Rift provided a lot of variations in terrain and environments, from rift valleys to stratovolcanoes (the iconic cone volcanoes, like Mount Kilimanjaro) and hence a great diversity of ecological niches.

In the end, the climatic cooling and changing terrain caused a transition from rain forest to savannah in large parts of Africa, which is where much of the final separation of hominids occurred over the last 15 million years. In short, these changes drove our ancestors out of the trees to the ground, where they could use their thumbs for something other than hanging on branches. Although this development was localized to Africa, the closing of the Tethys sea between Africa and Eurasia allowed land bridges and access for the spread of African mammals, including hominids, to Asia.

The branching of chimpanzees and humans probably occurred (by latest estimates) around 7 million years ago, as evident in the fossil remains of our oldest known ancestor, *Sahelanthropus tchadensis*, found in Chad, Africa. The branching off of humans or the genus *Homo* (of which the present human is the only species remaining) was most notably marked by consistent bipedalism, walking on two feet. The great apes are also able to walk like us, but they don't do it consistently. Standing is actually an odd position since it's unstable; without our toes and feet to constantly balance us, we would fall over. Four-legged animals are, of course, much closer to a stable position already. Moreover, running on two legs is in most cases slower than running on four legs, in which the entire torso in addition to the legs can be used in the stride. What, then, is the advantage of standing and walking on two feet?

There are many hypotheses about the origin of bipedalism, some of which have fared better than others. First, being able to carry food in your hands, in addition to your mouth, is a clear advantage. Being able to reach higher for food also seems like an advantage, but climbing for it is still better; so unless these apes lost all climbing ability, that notion doesn't seem to hold much water. Standing gives a better perspective for spotting predators and prey (though for the largely weaponless apes—with no sharp claws and fangs—predators were probably a bigger worry) and being able to do so while walking and running was probably an advantage (lots of animals can stand and get a better view, like the meerkat and bear). Looking bigger is another advantage in defending oneself and establishing dominance in social hierarchy and mating, and various animals including our cousin the gorilla do this.

Standing also gives better thermal regulation. It is easier for us than quadrupeds to cool off while running because standing increases exposure to air that helps evaporate sweat off the skin (and thus remove heat). Along the same lines, standing gives us better access to breezes, which are weaker closer to the ground. Indeed, the way humans cool is fairly unusual, although other primates as well as the horse also rely on sweat for cooling. Sweating is very efficient because it involves releasing water that then evaporates; the energy carried away by evaporation (turning water from liquid to gas) is huge (one of the biggest such energies of evaporation in nature) and so sweating is very effective. Other animals, like cats and dogs, achieve thermal regulation by panting, which is just exchanging heat with the atmosphere by bringing in cooler air and expelling warmer air. This is not as effective as sweating because air does not carry as much heat as does water vapor, which is one reason cats and (to a lesser extent) dogs are not great long-distance runners, certainly not compared to humans and horses. Modern humans rely on sweating off of naked skin, and our sweat is more watery and easier to evaporate than that of other sweating mammals, which have oilier sweat (which may be one of many reasons we lost our hair—to adapt to hotter environments). However, sweating works best in a warm and dry environment; if the air is hot and saturated with water, then our sweat won't leave our skin to go into the air (since water-saturated air can't take on more water vapor), which is why humans fare worse in high humidity (and why many of us prefer a dry heat to a humid heat). Although it may seem like I'm getting hung up on sweating, this mechanism is one of the clear examples that humans are ice age creatures. Our mode of thermoregulation would not have worked well 50 million (or even 30 million) years ago when humid, tropical, ice-free conditions covered the globe; our thermoregulation is suited for cooler and drier climates. Likewise, if human activity restores a global tropical environment, it won't match well with our dependence on sweating. It sounds like a marginal issue relative to, say, infectious diseases (which might also spread with global warming), but in fact the number one "extreme weather" killer is not the exotic hurricane or tornado, but heat waves.

Regardless of the cause of bipedalism, it freed up our ancestors' hands for holding tools with which to manipulate the environment, fare better against predators, and obtain food (from hunting to digging); thus, hands give an evolutionary advantage. At about 2.5 million years ago, there was a significant increase in the size of human brains. One current hypothesis is that the bigger brain resulted from a mutation that weakened the powerful jaw muscles anchored to the large sagittal crest on the top of most apes' skulls; this weakening allowed some freedom for the skull to evolve to a larger size. About this time, stone tool use appeared with Homo habilis, who had a larger brain than his predecessors. Stone tools were a big technological step with regard to enhanced hunting and digging as well as cutting and crushing food (to augment weaker jaws).

The controlled use of fire by Homo erectus followed, perhaps nearly 2 million years ago, although this is controversial; direct fossil evidence of ashes and burned bones in Africa and Eurasia certainly indicate that controlled fire arose within the last million years. Fire was a huge technological leap for several important reasons. First, it allowed humans to carry a heat source and expand into colder environmental niches than they could have otherwise. Second, it allowed cooking, which makes it much easier to consume many of the stringy proteins and tough plant cellulose molecules that would otherwise be very difficult to chew and digest. Fire also kills microbes and parasites in food. In the end, cooking created an evolutionary advantage by natural selection. Those who preferred raw meat were more likely to meet their demise, either by getting sick or not being able to eat as fast and as much, and therefore they would be removed from the gene pool. And fire is, of course, the precursor to much later technological advances like clearing crops, making pottery, forging better tools, and eventually powering machines.

The advent of fire, however, was perhaps the last major technological advance before the rise of agriculture hundreds of thousands of years later. In addition to Homo erectus, who vanished by at least 50,000 years ago (perhaps even earlier), both Neanderthal and Homo sapiens (that is, modern humans) appeared around 200,000 years ago. Neanderthal spread mostly into Europe and western Asia, using fire for colder climates, but the species went extinct by about 30,000 years ago, if not sooner. Perhaps both Neanderthal and Homo erectus vanished in competition with Homo sapiens, or perhaps through assimilation (since modern humans appear to have a small fraction of Neanderthal genes). However, a lot of other giant mammals went extinct at the hands of humans around the same time. Humans are still quite good at this.

The technological advances of tools and fire present an interesting evolutionary step in that they allowed adaptability to changing environments without major evolutionary changes to human physiology by natural selection. Of course, some modest natural selection still occurs anyway, for example, causing variations in the ultraviolet-absorbing melanin (aka pigment) in skin depending on Sun exposure. But in a larger sense, fire allowed humans to expand into harsher climates without, for example, evolving to have thicker fur or store much more fat. Thus, to some extent these technological developments were the first to allow humans to circumvent natural selection. However, this circumvention probably reached its peak with the advent of modern medicine.

On top of the final climate-cooling trend in the last 15 million years, there have been ups and downs of shorter glacial and interglacial (warmer) periods. The last long glacial period was the Pleistocene, which went from 2.6 million years ago to about 12,000 years ago, although there were, within this epoch, even shorter interglacial periods of a few thousand years each. During the Pleistocene the northern ice sheets extended far south in North America and Eurasia—for example, well into the midwestern United States and to the southern end of New York State; Long Island is the terminus or edge of where the glaciers dumped their deposits scraped off along the way from Canada. During some of this time, *Homo erectus*, Neanderthal, and *Homo sapiens* existed simultaneously, using the technology available to survive. Because of the large ice sheets on land, sea level was also lower, facilitating human migrations across connected landmasses, such as East Asia and North America along the Bering Strait as well as the British Isles and Europe. By the time the ice age ended, only *Homo sapiens* were left and they had already spread across most major continents.

The warming temperatures, starting around 12,000 years ago and peaking around 5,000 years later, coincided with the rise of agriculture in various forms. Warmer climates create better growing conditions not just because of higher temperatures but also by increasing the hydrological (evaporation and precipitation) cycle that provides fresh water and drives the cycling and accessibility of chemicals that go into the building blocks of life. Of course, biological production would occur in warmer climates without humans, so the important step is that by the time the warming happened, humans had the basic technology to domesticate some plants and animals, and to grow and store food as needed. For example, fire—which was harnessed during colder conditions—could be used for clearing forest to grow crops. Sharp stone tools could be used for tilling soil. (Metalworking wouldn't arise until the Bronze Age, around 5,000 years ago.) The fundamental crops for most societies were grains like wheat (and its various forms), domesticated in the Middle East (or Levant, including the Fertile Crescent), rice in East Asia, and maize (corn) in the Americas. Because agricultural societies were stationary and benefited by expanded use of land, their population growth was less constrained than for hunter-gatherers and herders whose nomadism would not support large populations. Agriculture also promoted territorialism and hence military protection of territories. In the end, it's not hard to see that the imbalances between agricultural societies and hunter-gatherers, along with likely collisions regarding territorial use (that is, nomadic versus domesticated use), could never end well for the hunter-gatherer societies.

The dominance and rise of agriculture also may have had an early anthropogenic effect on climate (which is called the Ruddiman hypothesis, named after its originator, geologist and paleoclimatologist William Ruddiman) long before the burning of fossil fuels. Clear-cutting and burning forests for crops (probably requiring even more land use per capita than today, given rudimentary technologies at that time) would emit carbon dioxide that would not be reabsorbed by less massive grain fields. Even the growth of rice in Asia, which peaked around 7,000 years ago, would increase the emission of methane (since rice paddies are marshes that undergo prolific decay and produce "swamp gas"), which is a very potent greenhouse gas. In principle, the warming 12,000 to 8,000 years ago should have been a short interglacial period, and Earth should have entered another ice age afterward. However, agricultural release of greenhouse gases circumvented that for thousands of years, and the burning of fossil fuels during the industrial age has continued that trend with a vengeance.

With the dominance of agricultural societies would come social hierarchies of workers (from artisans to field laborers) and rulers. In particular, the need for expansive infrastructure and organization to manage and protect resources (for example, irrigation and control of water sources, and granaries for food storage) would require military and political structures as well as writing for records, communication, and trade. Indeed, the oral and written recording of historical events and technical knowledge would provide an evolutionary advantage because it was information that would transcend the human lifespan and would give the species a leg up (helping it to avoid mistakes, such as during famine or flooding, made by prior generations). Taken together, these developments led to the dawn of civilization and history, which arguably began 7,000 years ago in Sumeria in southern Mesopotamia (modern Iraq).

One climatic event that may have triggered both the rise of Mesopotamian civilization and a large spread of cultures in western and Central Eurasia was the flooding of the Black Sea, also about 7,000 years ago. Melting of the Eurasian ice sheet at the end of the Pleistocene likely filled the Mediterranean, but warming conditions caused slow evaporation of the Black Sea, which was up till then freshwater and supported a large number of coastal cultures. The offset in sea level between the Mediterranean and Black Sea was as much as 140 meters, and the different water levels eventually incised a channel in the Bosporus Strait that let the Mediterranean slowly leak into the Black Sea, making it become the saltwater body it is today. Complete flooding would have taken about 3 years, but the gently sloping shoreline would have caused the water level to creep up relatively rapidly and render surrounding farmland useless. This event would have driven out the surrounding cultures, which radiated out in all directions, from the Mideast to Central Asia to western Europe, thus driving the migration of Indo-Europeans, Semitic tribes, and even the Ubaid into Mesopotamia to start the first Sumerian settlements. That the different cultures would all share a catastrophic flood history would also explain the common flood legends, including Noah's Flood in the Bible, the Utnapishtim legend in the Epic of Gilgamesh, and the Greek myth of Deucalion.

I also can't help but raise one fascinating and controversial notion that appeals especially to your geophysicist author. It's an idea championed by Jared Diamond and best summarized in his famous and elegant book, *Guns, Germs and Steel*. Diamond asks why colonial expansion was so one-sided in modern history. Europeans colonized other landmasses and subjugated or wiped out (often through disease) other cultures across the globe. Diamond's hypothesis is that continental orientation is the culprit. (Of course, how can I resist the idea that plate tectonics drove the tides of history as it has everything else since Earth formed?)

By the time humans had more or less spread everywhere, as Diamond's theory goes, the most important difference between cultures on each continent was not really the people on them but how their continents were aligned. Cultures in Eurasia (from East Asia to Europe) had a massive territory over which they could expand east or west and thus stay mostly in the same climatic zones. Expanding within a climate zone facilitates the transport of both crops and domesticated livestock without disrupting their growing and living conditions. However, such expansion is effective only over territory large enough that changes between different microclimates (for example, mountain ranges) have a small influence. That is, the climate zone is more or less uniform over hundreds and thousands of kilometers, but not over tens of kilometers, within which there might be a high desert range or a damp river valley.

On the whole, the Eurasian tectonic or continental setting provided much greater territory for expansion and diversification of agricultural populations and their technologies. In contrast, the other major continents are almost all aligned along a north-south orientation; thus, east-west migration along a constant climate zone was limited, and moving north or south would take crops and livestock out of their habitable zone. Thus, agricultural expansion was highly limited (and hunting-gathering might even have been more advantageous).

The extensive Eurasian expansion also led to domestication of more species of animals than on other continents, leading to greater exposure and hence immunities to various diseases. When the acquisitive Eurasian cultures expanded beyond their continent, they brought centuries of territorial military technology as well as diseases for which other continents' cultures had few defenses or immunities. In this way, small expeditionary forces could vanquish entire kingdoms, like that of the Spaniard Francisco Pizarro, who conquered the Incas in Peru with the vanguard of a devastating smallpox pandemic triggered by earlier Spanish settlers in the Caribbean.

I have no intention of going into the densely recorded past 7,000 years of human history. It's worth noting that those 7,000 years, about which so much has been written, account for only a half of one-millionth of the period covered in our story, starting 14 billion years ago. To revisit an analogy I introduced at the start of this book, if the Universe's history were sped up so you could watch it in one full 24-hour day, human history would zip by in 4/100 of a second, literally the blink of an eye. If we had allotted words in this book according to the length of time represented by each chapter, then human history would have been the final punctuation mark.

But in that incredibly brief period of time, the rise of humans and their unparalleled ability to manipulate the environment has a resulted in a species without predator, other than the life form that has been on this planet for almost 4 billion years—namely, bacteria. Without a serious competitor, our species has spread and the population has exploded faster than exponentially. In the last 2 centuries, we have found ways to exploit fossil energy sources hidden for hundreds of millions of years below the surface. These massive quantities of cheap energy have allowed a sudden jump in technology from which all of humanity (albeit some parts more than others) benefits today: from transportation to global communication to food production to medicine.

The downside of all this abundant cheap energy, in the form of environmental damage and unusual climate change, is apparently still too abstract to outweigh its enormous merits and potential, and so it might be a while before we change our habits. In the meantime, technology and medicine have shielded us from billions of years of natural selection (more so in developed countries). That means if things go south and we run out of resources to support this shield, a massive amount of our bloated population will, in effect, be chum for the microbes that have been patiently waiting (sorry to be so blunt and Vonnegutian). But profligate use of all resources is basically what unchallenged organisms do best. Bacteria alone in a petri dish consume food and energy until they're basically gone, and then that's it— the party's over.

But in the end, I like to think we are different from bacteria in a petri dish. Of all the other things humans have done, good and bad, we have developed ways of accumulating knowledge from which we can construct more knowledge. With language, history, and science, we are the first organisms on this planet (or anywhere we know of yet) that can make an educated guess about what happens next, beyond our own immediate future. Thus, we have the capacity and potential to act preemptively rather than reactively—or, worse yet, when it's too late. We will soon see, in this or the next generation, whether we can step up to the plate and use all our knowledge for the survival of generations yet unborn. If we do, it will be a unique moment in the history of life and, perhaps, even the Universe.

further reading

As I mentioned in the preface, the only unique things about this book are its brevity and, perhaps, the author's perspective. More comprehensive books exist that cover much, although not all, of the same material. Three excellent ways to continue are:

Jastrow, Robert, and Michael Rampino. *Origins of Life in the Universe.* Cambridge: Cambridge University Press, 2008.
Langmuire, Charles H., and Wally Broecker. *How to Build a Habitable Planet.* Rev. and expanded ed. Princeton, NJ: Princeton University Press, 2012.
MacDougall, J. D. *A Short History of Planet Earth: Mountains, Mammals, Fire and Ice.* Hoboken, NJ: Wiley & Sons, 1998.

Readers (especially my scientific colleagues) will notice I do not cite every possible source of information and discovery, or else the reference list would be longer than the text itself (and then I'd have to change the book's title, and my publisher is, for some reason, unusually fond of the title). The more well-known topics and heroic figures I mention are readily found in scholarly texts and popular science books. But for some overview, I suggest some general reading below. For more recent or more esoteric topics or results, I have tried my best to note the original authors and scientists, and provide more specific reading, also below. Thus for particular chapters, I list both general and specific reading.

1. UNIVERSE AND GALAXIES

GENERAL READING

Peebles, P. J. E., D. N. Schramm, E. L. Turner, and R. G. Kron. "Evolution of the Universe." *Scientific American*, October 1994, 50–57.
Singh, Simon. *The Big Bang: The Origin of the Universe.* New York: HarperCollins, 2005.

Trefil, James. *The Moment of Creation*. New York: Macmillan, 1983.

Turner, Michael. "Origin of the Universe." *Scientific American*, September 2009, 36–43.

SPECIFIC READING

Bromm, Volker, and Naoki Yoshida. "The First Galaxies." *Annual Review of Astronomy and Astrophysics* 49 (2011): 373–407.

Frieman, J. A., M. S. Turner, and D. Huterer. "Dark Energy and the Accelerating Universe." *Annual Review of Astronomy and Astrophysics* 46 (2008): 385–432.

Greene, Brian. "How the Higgs Boson Was Found." *Smithsonian Magazine*, July 2013. http://www.smithsonianmag.com/science-nature/ how-the-higgs-boson-was-found-4723520/.

Guth, A. H., and P. J. Steinhardt. "The Inflationary Universe." *Scientific American*, May 1984, 116–28.

Spergel, David N. "The Dark Side of Cosmology: Dark Matter and Dark Energy." *Science* 347, no. 6226 (2015): 1100–1102.

2. STARS AND ELEMENTS

GENERAL READING

Kirshner, Robert P. "The Earth's Elements." *Scientific American*, October 19, 1994, 58–65.

Lang, Kenneth R. *The Life and Death of Stars*. Cambridge: Cambridge University Press, 2013.

Young, Erick T. "Cloudy with a Chance of Stars." *Scientific American*, February 21, 2010, 34–41

SPECIFIC READING

Kaufmann, William J., III. *Black Holes and Warped Spacetime*. New York: W. H. Freeman, 1979.

Truran, J. W. "Nucleosynthesis." *Annual Review of Nuclear and Particle Science* 34, no. 1 (1984): 53–97.

3. SOLAR SYSTEM AND PLANETS

GENERAL READING

Elkins-Tanton, Linda T. *The Solar System*. 6 vols. New York: Facts on File, 2010.

Lin, Douglas N. C. "Genesis of Planets." *Scientific American*, May 2008, 50–59.

Lissauer, Jack J. "Planet Formation." *Annual Review of Astronomy and Astrophysics* 31 (1993): 129–74.

Wetherill, George. "Formation of the Earth." *Annual Review of Earth and Planetary Sciences* 18 (1990): 205–56.

SPECIFIC READING

Armitage, Phillip J. *Astrophysics of Planet Formation.* Cambridge: Cambridge University Press, 2010.

Canup, Robin M. "Dynamics of Lunar Formation." *Annual Review of Astronomy and Astrophysics* 42 (2004): 44175. doi: 10.1146/annurev.astro.41.082201.113457.

Chiang, E., and A. N. Youdin. "Forming Planetesimals in Solar and Extrasolar Nebulae." *Annual Review of Earth and Planetary Sciences* 38 (2008): 493–522.

Gomes, R., H. F. Levison, K. Tsiganis, and A. Morbidelli. "Origin of the Cataclysmic Late Heavy Bombardment Period of the Terrestrial Planets." *Nature* 435 (2005): 466–69.

Levison, H. F., A. Morbidelli, R. Gomes, and D. Backman. "Planet Migration in Planetesimal Disks." In *Protostars and Planets V*, ed. B. Reipurth, D. Jewitt, and K. Keil, 669–84. Tucson: University of Arizona Press, 2007.

4. CONTINENTS AND EARTH'S INTERIOR

GENERAL READING

Brown, G. C., and A. E. Mussett. *The Inaccessible Earth.* London: Chapman & Hall, 1993.

Condie, Kent C. *Plate Tectonics and Crustal Evolution.* Oxford: Pergamon, 1993.

"Our Ever Changing Earth." Special issue, *Scientific American* 15, no. 2 (2005).

Schubert, G., D. Turcotte, and P. Olson. *Mantle Convection in the Earth and Planets.* Cambridge: Cambridge University Press, 2001.

Stevenson, D. J., ed. *Treatise on Geophysics.* Vol. 9 of *Evolution of the Earth*, 2nd ed., ed. G. Schubert. New York: Elsevier, 2015.

Vogel, Shawna. *Naked Earth: The New Geophysics.* New York: Plume, 1996.

SPECIFIC READING

Bercovici, D. "Mantle Convection." In *Encyclopedia of Solid Earth Geophysics*, ed. H. K. Gupta, 832–851. Dordrecht, Netherlands: Springer, 2011.

Elkins-Tanton, L. T. "Magma Oceans in the Inner Solar System." *Annual Review of Earth and Planetary Sciences* 40 (2012): 113–39.

England, P., P. Molnar, and F. Richter. "John Perry's Neglected Critique of Kelvin's Age for the Earth: A Missed Opportunity in Geodynamics." *GSA Today* 17, no. 1 (2007): 4–9.

Glatzmaier, Gary A., and Peter Olson. "Probing the Geodynamo." *Scientific American*, April 2005, 50–57.

Stacey, F. D. "Kelvin's Age of the Earth Paradox Revisited." *Journal of Geophysical Research: Solid Earth* 105, no. B6 (2000): 13155–58.

5. OCEANS AND ATMOSPHERE

GENERAL READING

Allègre, Claude J., and Stephen H. Schneider. "The Evolution of the Earth." *Scientific American*, October 1994, 66–75.

Holland, H. D. *The Chemical Evolution of the Atmosphere and Oceans.* Princeton, NJ: Princeton University Press, 1984.

Kasting, J. F. "The Origins of Water on Earth." In "New Light on the Solar System," special issue, *Scientific American* 13, no. 3 (2003): 28–33.

SPECIFIC READING

Elkins-Tanton, L. T. "Formation of Early Water Oceans on Rocky Planets." *Astrophysics and Space Science* 302, no. 2 (2011): 359. doi: 10.1007/s10509-010-0535-3.

Valley, John W. "A Cool Early Earth?" *Scientific American*, October 2005, 58–65.

6. CLIMATE AND HABITABILITY

GENERAL READING

Bender, Michael L. *Paleoclimate.* Princeton, NJ: Princeton University Press, 2013.

Falkowski, P., R. J. Scholes, E. Boyle, J. Canadell, D. Canfield, J. Elser, N. Gruber, K. Hibbard, P. Högberg, S. Linder, F. T. Mackenzie, B. Moore III, T. Pedersen, Y. Rosenthal, S. Seitzinger, V. Smetacek, and W. Steffen. "The Global Carbon Cycle: A Test of Our Knowledge of Earth as a System." *Science* 290 (2000): 291–96.

Gonzalez, G., D. Brownlee, and P. D. Ward. "Refuges for Life in a Hostile Universe." *Scientific American*, October 2001, 60–67.

Kasting, J. F., and D. Catling. "Evolution of a Habitable Planet." *Annual Review of Astronomy and Astrophysics* 41 (2003): 429–63.

Ward, P. D., and D. Brownlee. *Rare Earth: Why Complex Life Is Uncommon in the Universe.* New York: Copernicus (Springer-Verlag), 2000.

SPECIFIC READING

Berner, Robert A. *The Phanerozoic Carbon Cycle.* Oxford: Oxford University Press, 2004.

Berner, R. A., A. C. Lasaga, and R. M. Garrels. "The Carbonate-Silicate Geochemical Cycle and Its Effect on Atmospheric Carbon Dioxide over the Past 100 Million Years." *American Journal of Science* 283, no. 7 (1983): 641–83.

Hoffman, Paul F., and Daniel P. Schrag. "Snowball Earth." *Scientific American,* January 2000, 68–75.

Huybers, P., and C. Langmuir. "Feedback Between Deglaciation, Volcanism, and Atmospheric CO_2." *Earth and Planetary Science Letters* 286, nos. 3–4 (2009): 479–91.

Raymo, M. E., and W. F. Ruddiman. "Tectonic Forcing of Late Cenozoic Climate." *Nature* 359, no. 6391 (1992): 117–22.

Walker, J., P. Hayes, and J. Kasting. "A Negative Feedback Mechanism for the Long-Term Stabilization of Earth's Surface Temperature." *Journal of Geophysical Research* 86 (1981): 9776–82.

7. LIFE

GENERAL READING

Clark, W. R. *Sex and the Origins of Death.* Oxford: Oxford University Press, 1996.

Hazen, R. M. *The Story of Earth.* New York: Viking, 2012.

Lane, N. *Life Ascending: The Ten Great Inventions of Evolution.* New York: Norton, 2009.

Orgel, L. "The Origin of Life on the Earth." *Scientific American,* October 1994, 76–83.

Ricardo, A., and J. W. Szostak. "Origin of Life on Earth." *Scientific American,* September 2009, 54–61.

Ward, P. D., and D. Brownlee. *Rare Earth: Why Complex Life Is Uncommon in the Universe.* New York: Copernicus (Springer-Verlag), 2000.

SPECIFIC READING

Corliss, J. B., J. Dymond, L. I. Gordon, J. M. Edmond, R. P. von Herzen, R. D. Ballard, K. Green, D. Williams, A. Bainbridge, K. Crane, and T. H. van An-del. "Submarine Thermal Springs on the Galápagos Rift." *Science* 203, no. 4385 (1979): 1073–83.

Doolittle, W. F. "Uprooting the Tree of Life." *Scientific American,* February 2000, 90–95.

Falkowski, P., R. J. Scholes, E. Boyle, J. Canadell, D. Canfield, J. Elser, N. Gruber,

K. Hibbard, P. Högberg, S. Linder, F. T. Mackenzie, B. Moore III, T. Pedersen, Y. Rosenthal, S. Seitzinger, V. Smetacek, and W. Steffen. "The Global Carbon Cycle: A Test of Our Knowledge of Earth as a System." *Science* 290 (2000): 291–96.

Mansy, S. S., J. P. Schrum, M. Krishnamurthy, S. Tobe, D. A. Treco, and J. W. Szostak. "Template-Directed Synthesis of a Genetic Polymer in a Model Proto-cell." *Nature* 454, no. 7200 (2008): 122–25.

Patel, B. H., C. Percivalle, D. J. Ritson, C. D. Duffy, and J. D. Sutherland. "Common Origins of RNA, Protein and Lipid Precursors in a Cyanosulfidic Protometabolism." *Nature Chemistry* 7, no. 4 (2015): 301–7.

Powner, M. W., B. Gerland, and J. D. Sutherland. "Synthesis of Activated Pyrimidine Ribonucleotides in Prebiotically Plausible Conditions." *Nature* 459, no. 7244 (2009): 239–42.

8. HUMANS AND CIVILIZATION

GENERAL READING

Behrensmeyer, K. "The Geological Context of Human Evolution." *Annual Review of Earth and Planetary Sciences* 10 (1982): 39–60.

Jurmain, R., L. Kilgore, W. Trevathan, and R. L. Ciochon. *Introduction to Physical Anthropology*. 14th ed. Belmont, CA: Wadsworth, 2013.

Silcox, M. T. "Primate Origins and the Plesiadapiforms." *Nature Education Knowledge* 5, no. 3 (2014): 1–6.

Tatersall, I. "Once We Were Not Alone." *Scientific American*, January 2000, 56–62.

Wong, K. "An Ancestor to Call Our Own." In special issue on evolution, *Scientific American*, April 2006, 49–56.

SPECIFIC READING

Behrensmeyer, K. "Climate Change and Human Evolution." *Science* 311 (2006), 476.

deMenocal, P. B. "Climate Shocks." *Scientific American*, September 2014, 48–53.

Diamond, J. *Guns, Germs and Steel*. New York: Norton, 1999.

Fagan, B. *The Long Summer: How Climate Changed Civilization*. New York: Basic Books (Perseus), 2004.

Jablonski, Nina G. "The Naked Truth." *Scientific American*, February 2010, 42–49.

Ruddiman, W. F. "How Did Humans First Alter Global Climate?" *Scientific American*, March 2005, 46–53.

Ryan, W., and W. Pitman. *Noah's Flood: The New Scientific Discoveries about the Event That Changed History.* New York: Simon & Schuster, 2000.

Sherwood, S. C., and M. Huber. "An Adaptability Limit to Climate Change Due to Heat Stress." *Proceedings of the National Academy of Sciences* 107, no. 21 (2010): 9552–55.

Stedman, H. H., B. W. Kozyak, A. Nelson, D. M. Thesier, L. T. Su, D. W. Low, C. R. Bridges, J. B. Shrager, N. Minugh-Purvis, and M. A. Mitchel. "Myosin Gene Mutation Correlates with Anatomical Changes in the Human Lineage." *Nature* 428, no. 6981 (2004): 415–18.

Wood, B. "Welcome to the Family." *Scientific American*, September 2014, 43–47.

acknowledgments

This book would not exist were it not for a group of Yale students who, in 2008, persuaded me (against my better judgment) to teach a class on the small topic of "everything," and then the many subsequent students who suffered through offerings of the class itself. I won't name all of them. You know who you are. Honestly, thanks. It was entertaining and I certainly learned a lot (even if you didn't).

I owe a great deal to many friends and collaborators: our scientific discussions and arguments over the years surely helped drive my curiosity about all the topics covered in this book. But I owe a particular debt of gratitude to the handful of colleagues who reviewed the book, either officially or unofficially. Peter Driscoll and Courtney Warren generously provided extensive reviews, many comments, and above all some perspective outside my own expertise—from astronomy to biology to anthropology. My old friend and PhD adviser Jerry Schubert kindly read several chapters and offered both encouragement and his characteristically candid comments (giving me some flashbacks to my graduate school years). An anonymous reviewer provided helpful comments. But I'm particularly indebted to Norm Sleep, who gave extensive reviews at two stages; I was thrilled to have Norm involved since he is a walking encyclopedia and has one of the most brilliant and far-reaching minds I know.

I also owe many thanks (and apologies for bouts of ill temper) to my editor Joe Calamia. Joe's scientific expertise, enthusiasm, patience, and unfailing sense of humor got me to the finish line.

I'm especially thrilled to thank my daughters, Sarah and Hannah, not just for their enthusiasm for the book but also for their careful reviews. Both of them are also scientists (okay, it happens), so I benefited from their expertise as well as from the retribution they exacted for all my ruthless and sarcastic shredding of their papers over the years; there is nothing like revenge to inspire candor (especially if it's worth a good laugh).

Finally, I thank my wonderful wife, Julie (also a scientist! . . . go figure), for her many readings, her encouragement, and above all her infinite patience, for more than just this book. *Sine te non sum* (again).

index